AF356659

Emerging Technologies for the Energy Systems of the Future

Emerging Technologies for the Energy Systems of the Future

Editors

Amjad Anvari-Moghaddam
Vahid Vahidinasab
Behnam Mohammadi-Ivatloo
Reza Razzaghi
Fazel Mohammadi

MDPI • Basel • Beijing • Wuhan • Barcelona • Belgrade • Manchester • Tokyo • Cluj • Tianjin

Editors
Amjad Anvari-Moghaddam
Aalborg University
Denmark

Vahid Vahidinasab
Nottingham Trent University
UK

Behnam Mohammadi-Ivatloo
University of Tabriz
Iran

Reza Razzaghi
Monash University
Australia

Fazel Mohammadi
University of Windsor
Canada

Editorial Office
MDPI
St. Alban-Anlage 66
4052 Basel, Switzerland

This is a reprint of articles from the Special Issue published online in the open access journal *Inventions* (ISSN 2411-5134) (available at: http://www.mdpi.com).

For citation purposes, cite each article independently as indicated on the article page online and as indicated below:

LastName, A.A.; LastName, B.B.; LastName, C.C. Article Title. *Journal Name* **Year**, *Volume Number*, Page Range.

ISBN 978-3-0365-1559-5 (Hbk)
ISBN 978-3-0365-1560-1 (PDF)

Contents

About the Editors

Amjad Anvari-Moghaddam is an Associate Professor at the Department of Energy Technology, Aalborg University, where he is the coordinator responsible for the Integrated Energy Systems Laboratory. His research interests include planning, control and operation management of microgrids, renewable/hybrid power systems and integrated energy systems with appropriate market mechanisms. He has (co)authored more than 200 technical articles, 5 books and 14 book chapters in the field. Prof. Anvari-Moghaddam serves as the Associate Editor of the *IEEE Transactions on Power Systems, IEEE Access, IEEE Systems Journal, IEEE Open Access Journal of Power and Energy,* and *IEEE Power Engineering Letters.* He is the Vice-Chair of IEEE-PES Danish Chapter and serves as a Technical Committee Member of several IEEE PES/IES/PEL and CIGRE working groups. He was the recipient of 2020 DUO-India Fellowship Award in Control and Management of Microgrids and Smart Energy Networks; the DANIDA Research Fellowship grant from the Ministry of Foreign Affairs of Denmark in 2018 for conducting cutting-edge research in Integrated Energy Systems and Energy Hubs; the IEEE-CS Outstanding Leadership Award 2018 (Halifax, Nova Scotia, Canada); the 2017 IEEE-CS Outstanding Service Award (Exeter, UK).

Vahid Vahidinasab (Senior Member, IEEE) received his PhD degree in electrical engineering from the Iran University of Science and Technology, Tehran, Iran, in 2010. Since 2010, he has worked in the Department of Electrical Engineering, Shahid Beheshti University (SBU), as an Assistant Professor. He also founded and managed the Soha Smart Energy Systems Laboratory at SBU. He has demonstrated a consistent track record of attracting external funds and managed industrial projects and closely worked with 12 large and complex national/international projects. From 2011 to 2018, he held a number of leadership roles at SBU and the Niroo Research Institute. In 2018, he moved to Newcastle University, where he worked as a Senior Research Associate and managed the inteGRIDy as an EU Horizon 2020 Project and also worked with the EPSRC Active Building Centre (ABC). Most recently, in 2021, he moved to Nottingham Trent University where he is currently a Senior Lecturer and leads teaching and research in the area of power and energy systems. His work is funded by the EPSRC-UK, EU-H2020, industry partners, and local utilities. His research interest is oriented to different research and technology aspects of energy systems integration, integrated energy systems as well as electricity markets. More specifically, he works on smart grid, microgrid, and nanogrid design, operation and economics, and the application of machine learning and advanced optimization methods in power and energy system studies (modeling, forecasting, and optimization). Dr. Vahidinasab is a member of the IEEE Power and Energy Society (PES) and the IEEE Smart Grid Community. He is also a member of the Editorial Board and a Subject Editor of the IET Generation, Transmission & Distribution, and an Associate Editor of the IET Smart Grid and IEEE Access. He is also the Guest Editor-in-Chief of a Special Issue on "Power and Energy Systems Operation in Time of Pandemics: Lessons Learned from COVID-19 Lockdown" of the *International Journal of Electrical Power & Energy Systems.* He was considered as one of the outstanding reviewers of the *IEEE Transactions on Sustainable Energy,* in 2018, and the *IEEE Transactions on Power Systems,* in 2020.

Behnam Mohammadi-Ivatloo, PhD, is a faculty member at the University of Tabriz, Tabriz, Iran, where he is the head of the Smart Energy Systems Lab. Before joining the University of Tabriz, he was a research associate at the Institute for Sustainable Energy, Environment and Economy, University of Calgary, Canada. He obtained his MSc and PhD degrees in electrical engineering from the Sharif University of Technology, Tehran, Iran. He has a mix of high-level experience in research, teaching, administration and voluntary jobs at the national and international levels. He has been PI or CO-PI in 20 externally funded research projects. He has been a Senior Member of IEEE since 2017 and a member of the Governing Board of Iran Energy Association since 2013, where he was elected as President in 2019. He is serving as either Editor or Associate Editor for different journals such as *IEEE Transactions on Power Systems*, *IEEE Access*, *IET Smart Grid*, and *Sustainability*. He is included in the 2018 and 2019 Thomson Reuters' list of the top 1% most cited researchers. His main areas of interest are integrated energy systems, renewable energies, microgrid systems, and smart grids.

Reza Razzaghi received his PhD degree in electrical engineering from the Swiss Federal Institute of Technology of Lausanne (EPFL), Lausanne, Switzerland, in 2016. In 2017, he joined Monash University, Melbourne, Australia, where he is currently a Lecturer (Assistant Professor) with the Department of Electrical and Computer Systems Engineering. His research interests include power system protection and control, distributed energy resources, microgrids, and electromagnetic transients in power systems. He has been the recipient of the 2019 Best Paper Award of the IEEE Transactions on EMC and the 2013 Basil Papadias Best Paper Award from the IEEE PowerTech Conference.

Fazel Mohammadi received his Doctorate degree in Electrical Engineering from the University of Windsor, Windsor, ON, Canada. Dr. Mohammadi is the founder of the Power and Energy Systems Research Laboratory. He is a Senior Member of IEEE and an active member of the International Council on Large Electric Systems (CIGRE), the European Power Electronics and Drives Association, and the Institution of Engineering and Technology (IET). His research interests include power and energy systems, high voltage engineering, power electronics, and smart grids.

Editorial

Emerging Technologies for the Energy Systems of the Future

Amjad Anvari-Moghaddam [1,*], **Vahid Vahidinasab** [2], **Behnam Mohammadi-Ivatloo** [3], **Reza Razzaghi** [4] **and Fazel Mohammadi** [5]

[1] Department of Energy Technology, Aalborg University, 9220 Aalborg, Denmark
[2] Department of Engineering, School of Science and Technology, Nottingham Trent University, Nottingham NG11 8NS, UK; vahid.vahidinasab@ntu.ac.uk
[3] Faculty of Electrical and Computer Engineering, University of Tabriz, 5166616471 Tabriz, Iran; bmohammadi@tabrizu.ac.ir
[4] Department of Electrical Engineering and Computer Systems Engineering, Monash University, Clayton 3800, Australia; reza.razzaghi@monash.edu
[5] Department of Electrical and Computer Engineering, University of Windsor, Windsor, ON N9B 1K3, Canada; fazel@uwindsor.ca or fazel.mohammadi@ieee.org
* Correspondence: aam@et.aau.dk

Citation: Anvari-Moghaddam, A.; Vahidinasab, V.; Mohammadi-Ivatloo, B.; Razzaghi, R.; Mohammadi, F. Emerging Technologies for the Energy Systems of the Future. *Inventions* **2021**, 6, 23. https://doi.org/10.3390/inventions6020023

Received: 17 March 2021
Accepted: 25 March 2021
Published: 27 March 2021

Publisher's Note: MDPI stays neutral with regard to jurisdictional claims in published maps and institutional affiliations.

1. Introduction

The way the world gets its energy is undergoing a rapid transition, driven by both the increased urgency of decarbonizing energy systems and the plummeting costs of renewable energy technologies. The road to the future will not be easy, and indeed, new technologies, markets, architectures, infrastructures, actors, and business models should be developed, and major changes will be required in the regulation of the energy systems to further support new business models and new consumption patterns. Such a transition requires rethinking every single aspect of energy systems, starting from the way energy is produced and harvested to the way that we dispatch and use it. In this area, it is also imperative to understand how to control and manage existing and emerging technologies to enhance the energy economy and efficiency by active participation in different services required by energy networks.

2. Emerging Technologies for Modern Energy Systems

This Special Issue covers new advances in the emerging technologies for modern energy systems from technical and management perspectives. In this context, an integrated and systematic view on different energy systems from the local energy systems and islands to the national and multinational energy hubs is a necessity that should be studied. From the customer side, it is required to have more intelligent appliances and smart customer services. The customers are required to be provided with more useful information and control options. On the supply side, a key challenge for energy systems of the future is the increased penetration of renewable energy sources. Hence, new operation and planning tools are required for hosting renewable energy sources as much as possible. Considering different aspects involving future energy systems deployment, the Special Issue on "Emerging Technologies for the Energy Systems of the Future" was launched to allow the gathering of contributions in the planning, integration, and management of such systems. In total, 15 papers were submitted to this Special Issue, and out of them, 10 were selected for publication. The accepted articles in this Special Issue cover a variety of topics, ranging from energy forecasting and the techno-economic analysis of renewable-based energy systems [1–3] to their hybridization [4,5], control [6,7] and integration [8–10].

The authors in [1] investigated the impact of wind generation forecasts on the market electricity prices by developing the autoregressive moving average (ARMA), ARMA with exogenous inputs (ARMAX) and a nonlinear autoregressive neural network (NAR), and NAR with exogenous inputs (NARX) artificial neural network (ANN) models of two differently conditioned power systems (Portugal and Poland) with a set of distinguishing features. It has been noted that the impact of wind power variability on the market electricity prices is caused by the relatively low marginal cost of the energy from wind farms, as wind energy is not burdened with fuel costs.

In [2], the authors proposed a deep learning technique called bidirectional long short-term memory(Bi-LSTM) to properly forecast the short-term load in a rural microgrid in Sub-Saharan Africa and solve the problems related to the reviewed methods, such as convolutional neural networks (CNNs), deep autoencoders (DAEs), recurrent neural networks (RNNs), and deep belief neural networks (DBNNs). The proposed method showed reasonable performance in processing large volume and time series data and dealing with missing data and overfitting during the training phase.

The authors of [3] provided a techno-economic analysis of different renewable energy technologies, including wind, solar photovoltaic (PV), and hybrid PV–wind systems. The objective was based on maximizing the renewable energy sources (RES) fraction, with the levelized cost of electricity (LCOE) being less than or equal to the local grid tariff in the studied system (i.e., Gwanda, Zimbabwe). The results confirmed that the PV–wind hybrid system had the best economic benefits, represented by the net present value (NPV) and the payback period (PBP), and the best technical performance. The authors determined the optimal size of the hybrid system and found that the PBP would be less than four years.

Focusing on hybrid renewable energy and energy storage systems for island power systems, the authors in [4] considered the Greek island power system of Astypalaia as a case study and designed a hybrid system, including a wind energy system and a battery energy storage system. Considering different economic indicators, the optimal combination was obtained, and then stability analysis was done to check the technical feasibility and sustainability of the system.

G.V. Kumar et. al. in [5] reviewed the energy storage system (ESS) participation in the provision of the ancillary services in a microgrid. The ESSs had a number of advantages, such as supply-demand balancing, smoothing of the intermittent generation from the renewable energy sources (RESs), enhancing the power quality and reliability (PQR), and provision of ancillary services like voltage and frequency regulation for the optimal operation of microgrids. In this paper, the status quo, opportunities, challenges, and real-world applications of the ESS application in providing different services in the electric power industry were explained.

The authors in [6] provided a model for secondary voltage control of a 100% renewable electricity system by extracting reactive power from renewable power technologies. Active and reactive power controls are considered based on the grid codes of countries with high penetration levels of renewable energy technologies. Based on a short-circuit calculation and sensitivity analysis, a pilot bus was selected to implement the secondary voltage control, which was based on a PID controller. In order to measure the voltage magnitude of buses with minimum cost, the optimal placement of phasor measurement units (PMUs) was performed. The optimal voltage magnitudes at busbars were calculated to achieve minimum power losses using optimal power flow. Optimization and simulation processes were performed using DIgSILENT and MATLAB software applications on a 100% renewable 14-bus system.

The authors in [7] presented a study of the load frequency control of an islanded marine microgrid composed of wind, tidal, and wave generators. First, they analyzed the impact of different supplementary control schemes on the dynamic performance of the load frequency control. Then, a novel black widow optimization technique was devised to design the hybrid system controllers and tidal supplementary controller. The authors demonstrated that the contribution of the tidal supplementary controller to the load or generation variation was more effective than the wind blade pitch controller in different operating conditions.

In [8], the generalization of solar water heater systems in Morocco and their potential energy-saving, economic and environmental benefits were presented. In particular, the authors analyzed the thermal performances and economic benefits of direct thermosyphon solar water heaters (TSWHs) for residential requirements in Morocco. Through various studies, it was concluded that the large-scale rollout of TSWHs could supply up to 70% of thermal energy loads and could provide USD 1250 million in savings on the total energy bills. This will also have significant environmental benefits by reducing CO2 emissions.

With the aim of highlighting the potential of solar water heater installations in Morocco, the work done in [9] involved the comparison of active and passive solutions for energy efficiency in buildings. By leveraging TRNsys Studio, a numerical simulation model of solar water heater installations was created considering various hot water demand scenarios (low, standard, and high) in six climatic zones defined in the Moroccan thermal regulation of constructions. The savings made by applying the various solutions were categorized into two parts: energy and financial. A comparison of energy savings showed that solar water heater solutions were superior to those achieved by building envelope actions. Additionally, the financial analysis of the savings on the calculated bills showed that solar water heater solutions were attractive in nonsubsidized butane gas prices.

Finally, the authors of [10] presented a brief review of definitions, research, and challenges in the integrated management of natural resources such as food, energy, and water. In this regard, some facts about demand growth and exponential consumption in these three areas were described, with emphasis on presented statistics. Then, the most critical research published in this field was reviewed, considering that it took a decade or so before the original idea was introduced. The most important policymakers of this emerging concept, including committees and conferences, and finally significant challenges and opportunities to the implementation, along with future insights, were addressed.

Funding: This research received no external funding.

Conflicts of Interest: The authors declare no conflict of interest.

References

1. Rogus, R.; Castro, R.; Sołtysik, M. Comparative Analysis of Wind Energy Generation Forecasts in Poland and Portugal and Their Influence on the Electricity Exchange Prices. *Inventions* **2020**, *5*, 35. [CrossRef]
2. Moradzadeh, A.; Moayyed, H.; Zakeri, S.; Mohammadi-Ivatloo, B.; Aguiar, A. Deep Learning-Assisted Short-Term Load Forecasting for Sustainable Management of Energy in Microgrid. *Inventions* **2021**, *6*, 15. [CrossRef]
3. Al-Ghussain, L.; Samu, R.; Taylan, O.; Fahrioglu, M. Techno-Economic Comparative Analysis of Renewable Energy Systems: Case Study in Zimbabwe. *Inventions* **2020**, *5*, 27. [CrossRef]
4. Fiorentzis, K.; Katsigiannis, Y.; Karapidakis, E. Full-Scale Implementation of RES and Storage in an Island Energy System. *Inventions* **2020**, *5*, 52. [CrossRef]
5. Kumar, G.; Palanisamy, K. A Review of Energy Storage Participation for Ancillary Services in a Microgrid Environment. *Inventions* **2020**, *5*, 63. [CrossRef]
6. Abdalla, O.; Fayek, H.; Abdel Ghany, A. Secondary Voltage Control Application in a Smart Grid with 100% Renewables. *Inventions* **2020**, *5*, 37. [CrossRef]
7. Fayek, H.; Mohammadi-Ivatloo, B. Tidal Supplementary Control Schemes-Based Load Frequency Regulation of a Fully Sustainable Marine Microgrid. *Inventions* **2020**, *5*, 53. [CrossRef]
8. Gargab, F.; Allouhi, A.; Kousksou, T.; El-Houari, H.; Jamil, A.; Benbassou, A. A New Project for a Much More Diverse Moroccan Strategic Version: The Generalization of Solar Water Heater. *Inventions* **2021**, *6*, 2. [CrossRef]
9. Gargab, F.; Allouhi, A.; Kousksou, T.; El-Houari, H.; Jamil, A.; Benbassou, A. Energy Efficiency for Social Buildings in Morocco, Comparative (2E) Study: Active VS. Passive Solutions via TRNsys. *Inventions* **2021**, *6*, 4. [CrossRef]
10. Abdi, H.; Shahbazitabar, M.; Mohammadi-Ivatloo, B. Food, Energy and Water Nexus: A Brief Review of Definitions, Research, and Challenges. *Inventions* **2020**, *5*, 56. [CrossRef]

Article

Comparative Analysis of Wind Energy Generation Forecasts in Poland and Portugal and Their Influence on the Electricity Exchange Prices

Radomir Rogus [1], Rui Castro [2,*] and Maciej Sołtysik [3]

[1] Instituto Superior Técnico, Lisbon, Portugal, and Silesian University of Technology, 44-100 Gliwice, Poland; rogusradomir@gmail.com

[2] INESC-ID/IST, University of Lisbon, 1049-001 Lisbon, Portugal

[3] Institute of Projects and Analyses, 44-100 Gliwice, Poland; maciej.soltysik@ipa-instytut.pl

* Correspondence: rcastro@tecnico.ulisboa.pt

Received: 21 June 2020; Accepted: 24 July 2020; Published: 25 July 2020

Abstract: Currently, the privileged position of wind energy producers is being weakened by their enforced participation in the market on equal terms. This requires accurate production forecasting. The main aim of this study is to comparatively examine the wind generation forecasts in Poland and Portugal, as well as to verify their influence on the day-ahead market prices. The statistical analysis revealed significant deviations of the forecasted and actual wind production in both countries, which referred to the corresponding spot and balancing prices caused considerable financial losses by the wind energy suppliers. In this paper, the influence of the wind generation forecasts on the spot prices has been examined through developed the auto-regressive moving average (ARMA), ARMA with exogenous inputs (ARMAX) and non-linear auto-regressive neural network (NAR), NAR with exogenous inputs (NARX)artificial neural network (ANN) models. The results have shown that the usability of the information of forecasted wind generation is not unequivocal in models developed for spot prices in Poland, mainly because of the randomness and volatility of recorded wind generation forecasts. However, in the case of Portugal, the forecasted wind generation occurred to be a valuable input in spot prices models, which results in an improvement in the models' accuracy.

Keywords: electricity price forecasting (EPF); wind power forecasting (WPF); spot market; balancing market; ARMAX; NARX-ANN

---

1. Introduction

Sustainable development of contemporary power system towards systems based on renewable energy sources (RES) requires facing some problems. These problems are mainly related to the inherent difficulty in predicting the availability of RES at a given time. This problem is made more difficult with the time-variation of the load and volatility of energy prices on the wholesale markets. One of the key aspects of reliable operation of the power system is proper planning of the supply versus demand, which, due to the RES generation uncertainty, becomes a struggle for responsible entities. The inaccuracy, of for instance wind generation forecasts, entails consequences of two primary natures: (i) technical—divergence of the forecasts around planned RES generation which impedes the transmission system operators' (TSOs) ability to properly and accurately plan the fulfilment of actual demand by available power capacity; and (ii) economical—the uncertainty of the total forecasted wind power supply translates into the uncertainty of the day-ahead and balancing market prices, which can result in financial losses of the market participants.

This study will examine the economic aspect identified above. Wind power variability influences the wholesale market price. This considerable impact is caused by the relatively low marginal cost of the energy from wind farms, because wind energy is not burdened with fuels costs. Therefore, an injection of the wind energy, featured by the lowest unit cost, shifts the supply-demand curve to the right, which results in a reduction of the market prices.

In the literature, numerous publications can be found which support the hypothesis that the wind energy forecast is a valuable input in wholesale market price forecasting. Moreover, a wide range of mathematical models have been developed as well (e.g., [1–3])—however, their universalism has not been tested on divergent market environments, such as Polish and Portuguese ones.

The abovementioned concerns have led to the statement of several research problems which are addressed by this paper. The main contributions of the research are the following:

- The quality of overall wind power forecasts in Poland and Portugal, as well as distinguishing the most relevant features of the forecasting errors. To address this issue, an evaluation of the wind power prognoses will be made using statistical measures, which will give a quantitative picture of the accuracy, characteristics and tendencies observed. The comparison of obtained statistic description for both countries will allow to distinguish the main differences between the forecasts and point out the main sources of uncertainty.

- The cost of inaccurate wind power forecasting globally from the perspective of wind power producers in both countries. This is accomplished by using market and wind generation data available for both countries to calculate the financial loss from inaccurate wind power predictions. This will illustrate the range and importance of appropriate and accurate wind generation forecasts.

- The extent to which the information on expected wind generation improves the accuracy of the wholesale market prices forecasts in both examined countries. The verification of whether wind generation has an impact in predicting market prices requires developing mathematical models. For this study, the four models commonly used in dealing with spot market prices time series will be approached: ARMA—auto-regressive moving average; ARMAX—auto-regressive moving average with exogenous inputs; NAR—non-linear auto-regressive neural network and NARX—non-linear auto-regressive with exogenous inputs neural network. The modelling process will be carried out independently using data from Polish and Portuguese systems. This will evaluate if the inclusion of wind generation data improves the spot prices predictions in both countries, and examine if a given model can be successfully applied for the data coming from diametrically different power and market systems.

The current stage of wind energy development in Poland and Portugal is comparable. Citing the Portuguese TSO, at the end of 2019, the installed onshore wind power capacity accounted for about 5.2 GW, while, in Poland, according to the state's Energy Regulatory Office, 2019 ended with 5.9 GW of wind power connected to the power system. So far, none of the countries has offshore wind installed capacity. After years of substantial and steady growth, in both cases, a plateau has been observed from 2016 onwards, and increases of installed power year to year are reaching no more than several dozens of MW. With regard to offshore wind power, both in Poland and in Portugal, no energy has been generated as of the end of 2019. In Portugal, a pre-commercial 25 MW offshore wind park using an innovative floating technology is being currently installed in deep waters. Though in Poland no construction works have been initiated yet, the special zone on the Baltic sea has been delimited, in order to develop offshore wind power of over 7 GW in the coming years.

The main objective of the study is to quantitatively evaluate the applicability of selected electricity price forecasting time series models in different datasets corresponding to Poland and Portugal market electricity prices. Moreover, we have verified the usefulness of wind power forecast as supplementary explaining variables of the models.

In the literature review concerning this research, a wide range of studies concerning the predictions of wind energy as well as spot market prices has been found. Although, usually, in the collected publications, a focus was put only on a single case of particular power system/power market. To the best knowledge of the authors, a multi-subject study has not been performed to reveal whether a given forecasting methodology applies to the separate cases of different national market environments, such as in Poland and Portugal. Although the Polish and Portuguese wind power capacity in the system is similar in amount when considering the share of this kind of energy the differences are substantial.

The main innovation of the paper is the use of the same methodology to analyse the influence of wind power forecasts in the market electricity prices of two differently conditioned power systems (Portugal and Poland), with a set of distinguishing features. This is accomplished by using a single data source, therefore making the comparison results robust and fully comparable.

The paper is organised as follows. To introduce the reader to the context of the abovementioned problems, Section 2 is devoted to the state-of-the-art research on the topics of wind power and spot market prices forecasting, together with particular focus on their interrelation. An overview of Portuguese and Polish RES sector and energy exchange institutions is offered in Section 3. Moreover, a description of the methodologies applied in solving the research problems stated in the introduction, including the basic statistical measures for dealing with forecasting error analysis and fundamentals of time series models constructed in terms of this study is presented in Section 4. In Section 5, the study results will be displayed, beginning from the statistical description of the wind generation forecasts error for Portugal and Poland, and related potential penalties suffered by the wind power producer globally. Finally, the spot prices forecasting results will be presented for the developed time-series models, with their comparative evaluation. At the end of this paper, the most significant findings will be highlighted in Section 6.

2. State-of-the-Art Review

Contemporarily, when the electricity is traded on wholesale markets, the uncertainty regarding the amount of wind energy supplied to the system has a direct influence on the market prices. Every excess or deficiency of energy in the system has a consequence in the financial outcomes of energy suppliers/receivers [1]. This elaboration concerns the two crucial problems in sustaining modern power systems, being a subject of intensive research within last decades, which are wind power forecasting (WPF) and electricity price forecasting (EPF). The increasing share of wind energy in the power systems, together with the tendency for equalisation of wind power producers with all the remaining electricity market participants, leads to the enhancement of the interaction between the WPF and EPF.

The impact of increased wind power penetration in power grid on the market player's behaviour and their decisions has been studied by Usaola (2008), who estimated that the uncertainty of wind power supplied to the system leads to significant reduction of the energy producers revenues, reaching 10% of maximum obtainable revenue [1].

The same finding has been obtained by Crespo-Vasquez et al. (2017), who constructed a decision-making model of the windfarm owner which underlined the loss of incomes resulting from the uncertainty of wind power output—in this case, the losses constituting 4% of the total earnings [4].

The analysed publications indicate that the improvement of WPF is crucial from the perspective of increasing the effectiveness of market actions taken by its contributors [1,2,4]. It also should be pointed out that the uncertainty of the energy produced in a system has not only negative consequences. The presence of the day-ahead market, as well as the balancing market, may allow members to increase their revenues, who may speculate the prices and gain profits only from buying/selling the energy depending on the market or imbalance prices [1].

A comprehensive and condensed study on WPF methods has been delivered by Zhao (2011), which gives a view on numerous aspects related to this issue. Above all, two main approaches in WPF models can be distinguished: (i) physical: the models are created based on the measurements, technical data of the windmills and air parameters provided by weather prognoses; (ii) statistical: including the artificial intelligence, based on the statistical models [5].

Among the most important issues related to the WPF is its scalability, which has been pointed out by Rasheed et al. (2014). The author emphasises that, for two instances of the same weather conditions in the wind farm, the power output may differ significantly, which is caused by the complexity of the terrain on which the installation has been set up [6].

The fact that there are not universally optimal WPF models for all the windfarms has been highlighted by Banerjee et al. (2017), who demonstrated that the most appropriate modelling approach depends also on the evaluation criterion [7].

The achievable WPF errors entail the necessity of knowing the primary sources of uncertainty. In a paper published by Monforti et al. (2017) the authors have attempted to detect and weight the main factors affecting inaccurate wind power predictions. The lack of comprehensive knowledge of the technical parameters of the wind farms and insufficient information on the wind fields are mentioned as the two main sources of WPF inaccurateness [8].

The desire for knowing the market prices in advance with acceptable reliability has become the subject of many studies. A remarkable focus has been put on forecasting the spot market prices [9–11]. Bessec et al. (2014) distinguish three main modelling methods for predicting the energy prices on the market: equilibrium and game theory, simulation and time series forecasting methods [12]. The basic models for prediction of time series include smoothing methods like averaging, exponential smoothing (e.g., Brown's, Holt's and Winter's methods). However, the results of these models suffer relatively high prediction errors [9].

Throughout numerous scientific publications, most attention is paid to the autoregressive (AR), moving average (MA) and their combination—ARMA—models, which have had success in forecasting economic phenomena [13]. The ARMA model can be useful in stationary processes, which are characterised by a constant value of the average in throughout the time domain. Since a lot of processes indicate a presence of a trend, the ARMA model cannot be applied directly—this problem is solved by extension of the ARMA model by including integration (the I term in ARIMA), which transforms the process with trend into a stationary one (ARIMA). Another difficulty comes from the seasonality of some time series in a short time, which has been addressed by the seasonal (SARIMA) model.

In the age of computers, artificial neural networks (ANN), a branch of computational intelligence (CI), has become increasingly popular in all range of predictions, mainly because of their forecasting accuracy and availability of dedicated software. Neural networks can be used successfully in circumstances which prevent statistical methods being applied. One main advantage of ANN is its adaptability for complex, dynamic and nonlinear relations. On the other hand, Weron (2014) emphasises that the neural networks can be susceptible to unexpected, rapid changes in the process [14].

Kolmek and Navruz (2012) performed the parallel forecasting processes for the Turkish energy spot prices utilising the ARIMA and ANN model, with the learning dataset constituting historical observations for time-period of 342 days. The latter model was achieved a smaller error, which was concluded as the main advantage of this method for statistical techniques (15.60% MAPE for ARIMA vs. 14.15% MAPE for ANN, calculated for a time horizon of 1 week) [15].

In his explicit study, Weron (2014) [14] lists the strengths and weaknesses of the electricity spot prices forecast in terms of the aforementioned methods, which have often been used separately a particular modelling approach. The weaknesses of each model are reduced though combining the modelling tools through dedicated computer software. The hybrid EPF models often result in performance improvements [16,17].

The most recent studies devoted to WPF and EPF strongly affirm the continuous necessity for proper projections of the wind power supplied to the system and their helpfulness in making decisions concerning the electricity market prices. Maldonado et al. [18], in their attempt to systematise the literature review concerning WPF, concludes that the field of improvement is sought more often in the direction of ANNs, which brings more promising results. However, what was highlighted by Santhosh et al. in paper [19], is that even the most advanced techniques do not solve the problem of generalisation capability, and do not assure suitability of a given model for individual windfarms.

In the scope of the same study, Santhosh indicates the latest development of hybrid models and highlight the importance of probabilistic forecasting. The influence of RES generation forecast on the day-ahead market prices have been confirmed by the study by Frade et al. [20], which was conducted based on the Iberian market data.

3. Portuguese and Polish RES Sector and Energy Exchange Institutions

Portugal is a country characterised by a very high share of RES in the total energy mix. Poland, while representing a steady growth in this area, does not reach the Portuguese share by far. It should be stressed that direct comparison on the RES share is improper since, throughout the continent, countries have developed diverse and specific energy systems, depending mainly on available domestic resources. Other key factors determining the shape of energy systems are climatic conditions (e.g., temperature), as well as the maximum energy independence. Although Poland has seen a considerable increase of RES in total energy production, the dominant primary energy source is coal, which is one of the most CO_2-intensive electricity generation technologies. Contrary to Poland, Portugal's energy mix is diverse, with no single dominant energy production technology [21,22].

The continuation of the European Union's (EU's) plans for clean and sustainable energy requires the adequate and optimal utilisation of the wide range of RES technologies to make it not only efficient, but also profitable. Poland and Portugal are countries separated by a vast distance, which results in different climatic conditions and different available sources of renewable energy generation. Taking into account the average annual wind speed, in both countries, the obtained measures are similar [23]. On the other hand, the solar energy in Portugal has a much higher potential for utilisation, with measures of the yearly solar irradiation being 1800 kWh/m^2 in Portugal and around 1200 kWh/m^2 in Poland.

The Towarowa Giełda Energii (TGE) was established in 1999, as the Polish government recognised the need to have an institution managing the energy trading on the emerging liberalised competitive market. Half a year after commission, the spot market was launched for market participants. Currently, TGE is the only company licensed by the state for managing the energy exchange. During the two recent decades, TGE expanded its area of activity into the emissions market exchange and origin certificates trading platform. In 2008, TGE introduced the Commodity Forward Instruments market, which allows it to maintain relatively constant prices across a wider time-horizon, as well as to optimise the costs of sales and purchases of energy [24].

While the TGE activities are limited to the within the country borders, the Iberian Electricity Market (MIBEL) is an example of the energy market expanded to the regional scale. It is a result of joint efforts of the Spanish and Portuguese governments. This convergence, having its origins in 2001, allows every market participant to make deals with other agents from all over the Iberian Peninsula. The integration of the hitherto separated energy systems required a series of undertakings, including the harmonisation of the electric network, laws, and economic environment. In 2004, both parts of the Santiago Agreement declared the creation of two sub-institutions, responsible for different aspects of the proper operation of the system as a whole. These were the OMI-Polo Português (OMIP), responsible for the forward market and OMI-Polo Español (OMIE), which was brought to existence to manage the spot market. The period between the Santiago Agreement and the launch of the Iberian Market (1 July 2007) was influenced by many factors, including political changes.

4. Methods

The research made in this paper can be divided into two main segments—the first one will refer to the evaluation of the wind power forecasts published for both Poland and Portugal in the European Network of Transmission System Operators for Electricity (ENTSO-E) Transparency Platform (https://transparency.entsoe.eu/), while the second part will concern the applicability of the ENTSO-E wind power forecast in the prediction models of the spot market prices.

4.1. Wind Generation Forecast Analysis

The analysis of wind power forecasts has been made by combining day-ahead wind generation forecasts and actual data, which has published by the ENTSO-E platform in hourly time steps. The data range has been assumed to be the calendar year 2016, which resulted in 8760 observations.

We have used the most up-to-date data available at the time of the study. In order to have a dataset representing the course of the entire calendar year, at the time of performing this analysis the most recent one was 2016. Since that time, the installed capacity in both Portugal and Poland has not changed significantly (see Table 1) and the market environment remained unchanged. Therefore, we believe that subjecting updated data to the models would not affect the conclusions of the study.

Table 1. Wind power installed capacity in Portugal and Poland in GW (2016 and 2019).

GW	2016	2019
Portugal	5.1	5.2
Poland	5.75	5.9

The prediction error is expressed as the difference between the realisation of the variable and its corresponding forecasted value for a given instant t:

$$e_t = P_t - P_t^*$$
(1)

where e_t is the absolute error of the forecast and P_t, P_t^* represent the actual wind power in the system and its day-ahead prediction, respectively, expressed in MW.

The statistical analysis of the obtained forecasting error will cover the calculation of the basic statistical measures presented in Table 2 [13].

Table 2. Statistical measures for wind power forecast evaluation.

Statistical Measure	Mathematical Expression	
Average $\bar{e}$, MW	$\bar{e} = \frac{1}{N} \sum_{t=1}^{N} e_t$	(2)
Standard deviation σ, MW	$\sigma = \sqrt{\frac{1}{N-1} \sum_{t=1}^{N} (e_t - \bar{e_t})^2}$	(3)
Kurtosis K	$K = \frac{\frac{1}{N} \sum_{t=1}^{N} (e_t - \bar{e})^4}{\sigma^4} - 3$	(4)
Skewness S	$S = \frac{N \sum_{t=1}^{N} (e_t - e)^3}{(N-1)(N-2)\sigma^3}$	(5)
MAPE, %	$MAPE = \frac{1}{N} \sum_{t=1}^{N} \left| \frac{e_t}{P_t} \right| \cdot 100\%$	(6)

where: e_t—forecast error [MW], N—total number of observations, t—time instance, P_t—actual value.

Moreover, the financial losses caused by inaccurate production planning and suffered by the wind energy suppliers in both countries will be estimated based on the ENTSO-E database. The calculations will be made using the actual and forecasted wind generation data, as well as their time-corresponding spot and balancing market prices for the calendar year 2016, in hourly time steps. The losses have been calculated assuming that the entire forecasted wind power is traded on the spot market, and the divergence between the wind forecast and actual production is traded on the balancing market. For this study, it was assumed that wind energy producers do not have any forward contracts. Additionally, wind producers rely entirely on the day-ahead plan and do not participate in the intraday market.

At this point, it has to be underlined that there are two divergent approaches to imbalance pricing in Poland and Portugal. In Poland, there exists only one, uniform imbalance market price. In contrast, in Portugal, two imbalance prices can be distinguished: the lower and upper imbalance price, depending on the imbalance characteristics (deficit or surplus).

Following the mentioned dissimilarities, the calculation of the financial loss has to be defined separately for both countries. The total loss of production plan deviation will be estimated as follows:

$$L = \sum_{t=1}^{N} \left|E_t - E_t^*\right| * \left(I_t^U - M_t\right), \qquad for \; E_t - E_t^* < 0 \tag{7}$$

$$L = \sum_{t=1}^{N} \left|E_t - E_t^*\right| * \left(M_t - I_t^L\right), \qquad for \; E_t - E_t^* > 0 \tag{8}$$

where: L—financial loss, E_t, E_t^*—actual and planned wind energy, respectively, M_t—spot market price, I_t^U—upper imbalance price; I_t^L—lower imbalance price (for the Portuguese case); $I_t^U = I_t^L = I_t$—imbalance price (for the Polish case); N—total number of observations; t—time instance.

4.2. Spot Market Prices Forecasting Models

To comparatively examine the impact of the forecasted wind power on the spot market prices in both countries, two representative modelling approaches have been selected from a wide range found in the literature: statistical (represented by the ARMAX model) and heuristic (the NARX model). For the evaluation of the usability of the wind energy as an input variable to these models, the corresponding values without the wind power information have been evaluated as well (the ARMA and NAR models). The comparison of ARMA vs. ARMAX and NAR vs. NARX using the data from Poland and Portugal will allow observation of what extent the models with additional external input perform better in two individual cases of Polish and Portuguese power markets.

4.2.1. Persistence Model

Persistence models, also named the naive models, are characterised by their simplicity. The forecasted value of model takes the value of the last observation.

$$y_{t+1} = y_t \tag{9}$$

4.2.2. ARMA and ARMAX—Autoregressive Moving Average Models

Box and Jenkins (1970) introduced a step-by-step methodology for modelling and estimation of time series by using the autoregressive (AR) and moving average (MA) models [25]. On the other hand, the ARMAX model is the extension of the ARMA model, utilising inclusion of the additional variable $u(t)$ in the model.

$$\begin{aligned} y(t) + a_1 y(t-1) + \cdots + a_{na}(t - n_a) \\ = b_1 u(t-1) + \cdots + b_{nb}(t - n_b) + \varepsilon(t) \\ + c_1 \varepsilon(t-1) + \cdots + c_{nc} \varepsilon(t - n_c) \end{aligned} \tag{10}$$

where: a_{na}—coefficients of the AR part of the model;

$\quad c_{nc}$—coefficients of the MA part of the model;

$\quad b_{nb}$—coefficients of the X part of the model;

$\quad n_a, n_b, n_c$—polynomial orders;

$\quad y(t)$—past values of the modelled quantity (here spot prices);

$\quad u(t)$—past values of the external explaining variable (here wind power forecast).

If all the entities related to the external variable $u(t)$ were removed from Equation (10), the model becomes an ordinary ARMA model, relying only on the past values of the explained variable.

Successful adoption of the ARMAX model entails the necessity of fulfilling the particular requirements of the Box-Jenkins method, as well as following the 3-step procedure, which is: model identification, estimation of the parameters and model verification. To properly identify the model, the examined time series has to meet particular requirements. First of all, it has to be stationary (there is no trend observable and the variance is constant). If the stationarity is not confirmed, the time series has to be modified to achieve stationarity. Secondly, the parameters of the model have to be determined employing computational methods (e.g., the maximum likelihood estimation or non-linear least-squares estimation). Further, the estimated parameters have to be verified statistically to prove their significance to the model. In this particular case, the ARMAX model data requirements have been validated by the integration of the spot market prices time series. To remove the periodicity of the series, the spot prices series have been differenced by time-lag of 24 h.

The best model structure, i.e., polynomial orders, was obtained iteratively, using the looping capabilities of the software used.

4.2.3. NAR and NARX—Artificial Neural Network Models

The origins of artificial neural networks (ANN) are in 1943, when the first artificial neuron model was introduced by McCulloch and Pitts. Contemporarily, ANNs are commonly used in statistics and signal processing. Despite the strong development of this branch of science, the range of ANNs application is continuously expanding. The standard structure of the ANN consists of input units (vector of numbers provided by the user), hidden units (representing the intermediate calculations) and output units (vector of the model results) linked by particular weights. The weights reflect the importance of a given signal in the model—the higher the weight value, the more significant the input signal in the process of output determination. Therefore, the weights can take both negative or positive values [26].

The weights adjustment constitutes the learning process of the network, which, in the case of this study, has been made by use of the Levenberg-Marquardt algorithm, considered one of the most commonly used ANN learning techniques, and characterised by high effectiveness in feedforward networks training [27].

If the external signal is removed from the NARX model, it becomes the NAR model, with only the past values of spot market prices as the explanatory variables. The best model structure i.e., the number of hidden layers and number of input signals (here, the number of lags of time series) was obtained iteratively using the looping capabilities of the software used.

4.2.4. Forecasting Approaches

The data used in the models was the wind generation forecast and spot market prices recorded in hourly time steps for 2016, as shown in Figure 1 for Poland and Portugal.

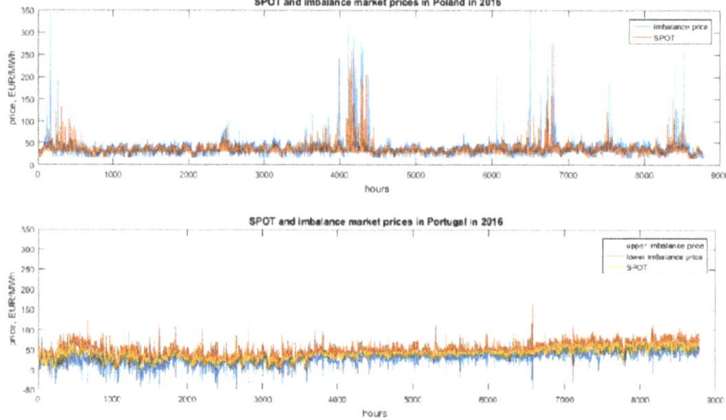

Figure 1. Spot and imbalance market prices in Poland and Portugal in 2016.

Comparing both spot prices series, Poland's energy price denotes high volatility, while the electric energy prices in Portugal are more clustered around a constant value. Moreover, a slightly increasing trend is noticeable in the Portuguese energy prices, especially in the second half of the year. Since the time series reveal periodicity, which eliminates the direct application of ARMA and ARMAX models, all forecasts will be made based on differenced (integrated) time series by 24 hours' lag, as already mentioned.

It was decided to divide the data into two subsets: (i) learning dataset—consisting of observations from January to November 2016, which was used for the estimation of the models; and (ii) testing dataset—December 2016, which was used for verification of the models' performance. Thus, the forecasts were made in a step-ahead manner for December 2016.

Moreover, the forecasting by all models was made by splitting the training data set into static and dynamic groups. The static approach uses a singular estimation of the model's parameters, which are used in an unchanged form for the entire prediction horizon (1 month). On the other hand, in the dynamic approach, the model's parameters are updated every prediction step, based on the modified learning dataset, which is shifted also by one step. As such, it was decided to test several cases which differed by the size of learning dataset. These were defined as follows:

- Static approach—fixed learning dataset—from January to November 2016 (case S1);
- Static approach—fixed learning dataset—November 2016 (case S2);
- Dynamic approach—"moving" learning dataset—preceding week (case D1);
- Dynamic approach—"moving" learning dataset—preceding 24 h (case D2).

The mean absolute percentage error (MAPE) has been chosen as the parameter to evaluate the quality of obtained forecasts, the has been selected and calculated for each of the abovementioned cases (Equation (6)), for Poland and Portugal, referred to the entire month of December 2016.

In order to have a perspective on the financial burden corresponding to the inaccuracy of the performed forecasting models, the total uncertainty of sales of the wind energy (SU) in December 2016 (744 h) has been calculated according to Equation (11), under the assumption that the entire amount of wind energy is traded on the spot market:

$$SU = MAPE \cdot \sum_{t=1}^{744} E_t^* \cdot M_t \tag{11}$$

where SU—total sales uncertainty (EUR), E_t^*—traded forecasted wind energy volume in hour t (MWh), M_t—spot market price in an hour t (EUR/MWh).

5. Results and Discussion

The wind power error value (expressed as the difference between the realisation of the variable and its corresponding forecasted value) has been calculated for each hour of 2016, and divided into ranges to present the error distribution on a histogram, as shown in Figure 2.

The main conclusion from the analysis of the above chart is that the wind generation forecasts are generally underestimated in both countries, since the result of subtracting the forecasted value from the actual value of power results in positive values, between 0 and 200 MW.

To get a deeper view of the characteristics of the error distribution, the statistical parameters of the examined dataset have been calculated. The statistical description has been carried out to quantitatively compare the inaccuracies of wind power forecasts for both countries, including the calculation of statistical measures, which are presented in Table 3 and are referred to in Section 3.

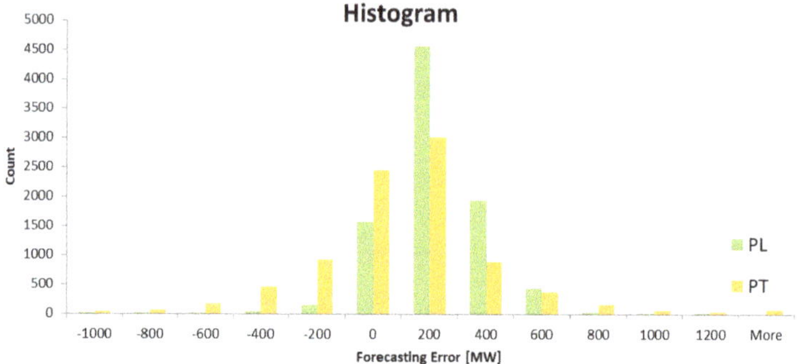

Figure 2. Aggregated wind power error forecast distribution for Poland and Portugal, 2016.

Table 3. Statistical description of the wind energy forecast error for Poland and Portugal in 2016.

Statistical Measure	Poland	Portugal
Average, MW	118.23	7.52
Standard deviation, MW	173.00	353.30
Kurtosis	7.17	13.02
Skewness	0.399	−0.307
Minimum, MW	−1189.00	−3659.00
Maximum, MW	1789.00	1934.00
MAPE, %	15.00	24.20

Analysing Table 3, one may conclude that both Polish and Portuguese forecasts are underestimated, because the average error values exceed 0 significantly, especially in Poland, reaching 118 MW. Although the average is much higher in Poland compared to Portugal, the Portuguese forecast is characterised by higher volatility, which is indicated by the values of kurtosis and standard deviation. Comparing the kurtosis of both forecasts' errors, the higher tendency for clustering of the error near the average value occurs in the forecast prepared for Poland. In both cases, the extreme values of the forecast error reached gigawatts. Comparing the MAPE values, the Polish forecasts perform considerably better, reaching 15% on average. The skewness values of examined samples took the opposite sign for Poland and Portugal, which means that the asymmetry of the error distributions has the opposite direction. Comparing the wind generation forecasts error (up to 2 GW) with the overall generation rates in the countries (reaching up to about 10 GW in Portugal and 50 GW in Poland) leads to the question in what extent these forecasts influence the total energy system and energy markets.

The potential financial losses have been calculated assuming that the entire forecasted wind power is traded on the spot market, and the divergence between the wind forecast and actual production is traded on the balancing market. The results of the performed calculations are presented in Table 4:

Table 4. Financial losses resulting from wind generation forecast deviation in 2016.

	Poland	Portugal
Total financial loss, EUR	1,900,592	20,812,563
Number of hours in 2016, when forecast deviation brought benefits	4510	814
Total volume of imbalanced energy, MWh	1,358,182	1,944,235
Unit cost of imbalanced energy, EUR/MWh	1.40	10.70

Although the total wind-generated imbalanced energy volume in 2016 is not significantly different between Poland and Portugal, the financial losses in the case of the Iberian country reached almost EUR 21 million, which was over ten times more than in the case of Poland. Although the yearly-aggregated forecast deviation brought losses, there were hours, when it resulted in positive income ($L < 0$). In Poland, more than half of hours in the year brought benefits from incorrect forecasts (4510 h in 8784 of total).

To directly compare the results, the unit costs of the forecast error was calculated by the division of the total imbalance by the total financial loss in 2016. As shown in Table 4, a 1 MWh planning error in Portugal resulted in EUR 10.7 of costs, which was almost 10 times larger when compared with Poland. Such a significant discrepancy comes from the imbalance pricing system, which is different in these countries. In Portugal, there are two levels of balancing market price, depending on the character of imbalance (surplus/deficiency). In Poland, there is only one universally valid imbalance price, which allows for market speculation. In the circumstance of a single, uniform imbalance price, there exists a considerable probability of obtaining income from selling energy on the balancing market instead of the spot market. The Polish wind energy seller may speculate on the difference between the spot and imbalance prices, which may encourage him to intentionally distort the production plan. In contrast, the Portuguese energy seller may expect only the negative outcomes from an inappropriate production plan.

As far as the influence of wind energy forecasts on spot prices is concerned, the forecasting results with the use of ARMA, ARMAX, NAR and NARX models are presented in Tables 5 and 6, respectively. Results where the inclusion of wind power forecasts led to better performance when compared to the case without wind power information are in bold text. The information of the key model parameters is also shown. Moreover, the uncertainty of the sale as defined in Equation (11) is also displayed.

Table 5. Comparison of the auto-regressive moving average (ARMA) and ARMA with exogenous inputs (ARMAX) spot prices forecasting models.

POLAND										
	ARMA				ARMAX					SU_{ARMA}-SU_{ARMAX}
CASE	PARAMETERS		MAPE	SU_{ARMA}	PARAMETERS			MAPE	SU_{ARMAX}	
	p	q		(EUR)	na	nb	nc		(EUR)	(EUR)
S1	1	0	6.98%	3,448,565	1	0	0	6.97%	3,443,624	-
S2	1	0	6.90%	3,409,040	1	0	0	6.89%	3,404,099	-
D1	**1**	**2**	**7.04%**	**3,478,209**	**2**	**2**	**0**	**6.79%**	**3,354,693**	**123,516**
D2	1	0	8.23%	4,066,145	1	0	1	6.81%	3,364,574	-

PORTUGAL										
	ARMA				ARMAX					SU_{ARMA}-SU_{ARMAX}
CASE	PARAMETERS		MAPE	SU_{ARMA}	PARAMETERS			MAPE	SU_{ARMAX}	
	p	q		(EUR)	na	nb	nc		(EUR)	(EUR)
S1	5	4	3.66%	2,262,710	5	4	5	3.57%	2,207,070	55,640
S2	4	4	3.71%	2,293,622	5	2	5	3.57%	2,207,070	86,552
D1	**3**	**3**	**4.24%**	**2,621,282**	**4**	**2**	**4**	**3.64%**	**2,250,346**	**370,936**
D2	3	3	6.09%	3,765,002	3	5	3	7.77%	4,803,623	-

To have a reference of the models constructed in this paper, the persistence model has been simulated as well. Applying Equation (9) to the validation dataset (December 2016), the persistence model has been constructed and evaluated by obtaining the value of MAPE, which took a value of 6.92% in Poland and 4.50% in Portugal.

The S1, S2 and D2 cases have not been highlighted, because in these examples the ARMAX model's *nb* coefficient was zero. This is the order of polynomial composed of external variable values, which means that the ARMAX model becomes ordinary ARMA model.

Analysis of Tables 5 and 6 show that when comparing the forecasting results globally, the Portuguese spot market prices are significantly more predictable than the Polish case using the models constructed for this study. Another finding is that usually, the dynamic cases perform worse than the static ones thus the model estimation should include wider ranges of data. Additionally, in the right-hand-side column, the difference between the sales uncertainty (SU) of the models with and without the inclusion of the wind generation forecast has been calculated. The SU value varied between EUR 55,000 and EUR 235,000, in the case of models based on Portuguese data.

Table 6. Comparison of the non-linear auto-regressive neural network (NAR) and NAR with exogenous inputs (NARX) spot prices forecasting models.

		POLAND								
	NAR				NARX					SU_{NAR}-SU_{NARX}
CASE	PARAMETERS		MAPE	SU_{NAR} (EUR)	PARAMETERS			MAPE	SU_{NARX} (EUR)	(EUR)
	delays	Hidden layers			delays	ex.Input delays	Hidden layers			
S1	1	5	6.71%	3,315,168	3	5	2	6.74%	3,329,990	-
S2	2	2	6.55%	3,236,118	5	5	2	6.63%	3,275,643	-
D1	2	3	7.54%	3,725,241	2	1	1	8.12%	4,011,798	-
D2	3	4	9.83%	4,856,647	4	4	4	8.78%	4,337,880	518,767
		PORTUGAL								
	NAR				NARX					SU_{NAR}-SU_{NARX}
CASE	PARAMETERS		MAPE	SU_{NAR} (EUR)	PARAMETERS			MAPE	SU_{NARX} (EUR)	(EUR)
	delays	Hidden layers			delays	ex.Input delays	Hidden layers			
S1	5	1	3.74%	2,312,169	5	2	5	3.63%	2,244,164	68,005
S2	4	5	3.71%	2,293,622	4	5	3	3.64%	2,250,346	43,276
D1	1	1	3.88%	2,398,720	5	5	5	3.67%	2,268,893	129,828
D2	3	4	4.58%	2,831,479	5	4	1	4.20%	2,596,553	234,926

Recalling the reference forecasts of the persistence model (6.92% MAPE in Poland and 4.50% in Portugal), it can be observed that in the case of Poland the obtained models rarely perform better than the persistence one, while, in Portugal, the MAPE obtained by time series models is often below 4%.

Comparing country-to-country, it can be observed that the wind generation forecast is valuable for predicting the spot prices in Portugal, because in seven out of eight cases, the "X" model returned more accurate results than its counterpart without information of wind generation forecast (ARMA vs. ARMAX and NAR vs. NARX). In the case of predicting the Polish spot market prices, only in two instances out of eight did the addition of wind production forecasts improve the model output. The cumulated simulation results show that wind generation forecasts influence the spot market prices in Portugal. In the case of the Polish market, has been shown that the forecasts do not have an impact on the spot market price. As a source of this difference, the authors consider the significant divergence in the wind generation share in the total electricity production. In the case of Portugal, the relative amount of the energy injected to the system from wind turbine generators is much higher, thus, may have a greater impact on the market behaviour.

As a general tendency, the Portuguese spot market prices were more predictable than the Polish time series. This is shown by the one-sided difference of the MAPE measure in all the cases.

Moreover, when comparing the results obtained from the four models for the individual learning dataset approaches (S1, S2, D1 and D2), a global result is that the models of "static" character, those being estimated on a fixed dataset, resulted in obtaining lower errors. As for the reason, the authors assume the instantaneous peaks of the prices occurring within a week or a day, which may significantly influence the estimation and impact the model's prediction accuracy. In the situation when the time range of the data was wider (1 month, 11 months), this impact was mitigated.

The results answer the fundamental question of this study because as it is observed, the impact of wind power on the market prices is not unequivocal. An application of a given model with wind power variable as input occurred to be more accurate in the case of Portugal, while in Poland not.

The outcome of the study demonstrates consistency with former research on the topic. As highlighted in the study by Maciejowska et al. [28], the RES generation forecasts published on the ENTSO-E platform are encumbered with systematic errors, which gives space for future improvement in EPF. The analysis of WPF and EPF for Poland and Portugal confirmed the relevance of the forecasting error, which translated to the level of financial loss or gain, strongly amplifying the uncertainty of sales. The calculated values of MAPE for ARMA and ANN models, compared to the ones reported in the studies devoted to spot market forecasting (e.g., Kolmek and Navruz, [15]), can be found satisfactory. However, direct comparison between studies in the literature should be performed with caution. Our parallel study addressed this deficiency by comparatively analyse the separate cases of two countries, but with a data feed completely correspondent to one another in time and scope. That allowed to unambiguously evaluate the accurateness of EPF models and their receptivity for external data input as the wind power forecast.

6. Conclusions

The analysis of the wind power forecasts for Poland and Portugal resulted in significant forecasting errors, based on calculations made by the use of data from the ENTSO-E platform. Large discrepancies have been observed for both analysed countries. Analysis of the wind forecast error distribution revealed that there exists a difference in the estimation of the wind generation forecast among the countries. In Portugal, the forecast often predicted lower values than the actual ones. In the case of Poland, these forecast values overestimated the production compared to actual values. This finding opens a new research opportunity which would investigate the application of systematic error correction in overall wind production forecasts.

Calculation of the financial losses from inaccurate wind power forecasts led to the conclusion that the uncertainty of the wind energy production forecasting results in remarkable potential income losses for the wind energy producers, reaching millions of EUR in global (national) scale. As shown in this study, in the Polish case, there exists a large possibility of market speculation, since, in many instances, the sales on the balancing market were more profitable than in the spot market.

Analysis of the spot price forecasting models shows that in the case of Poland, the applicability of wind power as the input to a particular model is uncertain. On the other hand, a decrease of MAPE values has been demonstrated when adding the wind power input variable to the models based on Portuguese data, which translated into lower values of sales uncertainty of the generated wind power. These opposite observations reveal that the advantage of models including wind power information is not universal among the individual countries.

Author Contributions: Conceptualisation, R.R.; methodology, R.R.; software, R.R.; validation, R.C. and M.S.; formal analysis, R.R. and R.C.; investigation, R.R.; resources, R.R., R.C. and M.S.; data curation, R.R.; writing—original draft preparation, R.R.; writing—review and editing, R.C.; visualisation, R.C. and M.S.; supervision, R.C. and M.S.; project administration, R.C.; funding acquisition, R.C. All authors have read and agreed to the published version of the manuscript.

Funding: This research was funded by Fundação para a Ciência e a Tecnologia (FCT), grant number UIDB/50021/2020.

Acknowledgments: The authors want to thank Matthew Gough for his careful revision of the English language of this manuscript.

References

1. Usaola, J. Participation of wind power in electricity markets. In *Preliminary Report from 6th FP European Project (Reference 38692) and IEMEL—Research Project of the Spanish Ministry of Education (Reference ENE2006-05192/ALT)*; Universidad Carlos III de Madrid: Madrid, Spain, 2008.
2. Jónsson, T.; Pinson, P.; Nielsen, H.A.; Madsen, H.; Nielsen, T.S. Forecasting electricity spot prices accounting for wind power predictions. *IEEE Trans. Sustain. Energy* **2013**, *4*, 210–218. [CrossRef]
3. Erni, D. Day-Ahead Electricity Spot Prices-Fundamental Modelling and the Role of Expected Wind Electricity Infeed at the European Energy Exchange. Ph.D. Thesis, University of St. Gallen, St. Gallen, Switzerland, 2012.

4. González-Aparicio, I.; Zucker, A. Impact of wind power uncertainty forecasting on the market integration of wind energy in Spain. *Appl. Energy* **2015**, *159*, 334–349. [CrossRef]

5. Zhao, X.; Wang, S.; Li, T. Review of Evaluation Criteria and Main Methods of Wind Power Forecasting. *Energy Procedia* **2011**, *12*, 761–769. [CrossRef]

6. Rasheed, A.; Suld, J.K.; Kvamstdal, T. A Multiscale Wind and Power Forecast System for Wind Farms. *Energy Procedia* **2014**, *53*, 290–299. [CrossRef]

7. Banerjee, A.; Tian, J.; Wang, S.; Gao, W. Weighted Evaluation of Wind Power Forecasting Models Using Evolutionary Optimization Algorithms. *Procedia Comput. Sci.* **2017**, *114*, 357–365. [CrossRef]

8. Monforti, F.; Gonzalez-Aparicio, I. Comparing the impact of uncertainties on technical and meteorological parameters in wind power time series modelling in the European Union. *Appl. Energy* **2017**, *206*, 439–450. [CrossRef]

9. Misiorek, A.; Weron, R. Forecasting spot electricity prices with time series models. In Proceedings of the The European Electricity Market International Conference (EEM-05), Lodz, Poland, 12–11 May 2005; pp. 133–141.

10. Voronin, S.; Partanen, J. Price forecasting in the day-ahead energy market by an iterative method with separate normal price and price spike frameworks. *Energies* **2013**, *11*, 5897–5920. [CrossRef]

11. Franco, J.; Blanch, E.; Ribeiro, C.O.; Rego, E.; Ivo, R. *Forecasting Day Ahead Electricity Price Using ARMA Methods*; Universidade de São Paulo: São Paulo, Brazil, 2015; Available online: https://upcommons.upc.edu/bitstream/handle/2117/82251/TFG2.pdf (accessed on 1 April 2018).

12. Bessec, M. Forecasting electricity spot prices using time-series models with a double temporal segmentation. In Proceedings of the 2nd International Symposium on Energy and Finance Issues (ISEFI-2014), Paris, France, 28 March 2014; p. 34.

13. Witkowska, D. *Podstawy Ekonometrii I Teorii Prognozowania*; Oficyna Ekonomiczna: Czestochowa, Poland, 2006; ISBN 83-7484-029-3.

14. Weron, R. Electricity price forecasting: A review of the state-of-the-art with a look into future. *Int. J. Forecast.* **2014**, *30*, 1030–1081. [CrossRef]

15. Kolmek, M.A.; Navruz, I. Forecasting of the day-ahead price in electricity balancing and settlement market of Turkey by using artificial neural networks. *Turk. J. Electr. Eng. Comput. Sci.* **2015**, *23*, 841–852. [CrossRef]

16. Tan, Z.; Zhang, J.; Wang, J.; Xu, J. Day-ahead electricity price forecasting using wavelet transform combined with ARIMA and GARCH models. *Appl. Energy* **2010**, *87*, 3606–3610. [CrossRef]

17. Shafie-khah, M.; Moghaddam, M.P.; Sheikh-El-Eslami, M.K. Price forecasting of day-ahead electricity markets using a hybrid forecast method. *Energy Convers Manag.* **2011**, *52*, 2165–2169. [CrossRef]

18. Maldonado, J.; Solano, J. Wind power forecasting: A systematic literature review. *Wind Eng.* 2019. [CrossRef]

19. Santhosh, M.; Venkaiah, C.; Kumar, D.M.V. Current advances and approaches in wind speed and wind power forecasting for improved renewable energy integration: A review. *Eng. Rep.* **2020**, *2*, e12178.

20. Frade, P.M.S.; Vieira-Costa, J.V.G.A.; Osório, G.J.; Santana, J.J.E.; Catalão, J.P.S. Influence of Wind Power on Intraday Electricity Spot Market: A Comparative Study Based on Real Data. *Energies* **2018**, *11*, 2974. [CrossRef]

21. International Energy Agency. *Energy Policies of IEA Countries—Portugal—2016 Review*; International Energy Agency: Paris, France, 2016.

22. International Energy Agency. *Energy Policies of IEA Countries—Poland —2016 Review*; International Energy Agency: Paris, France, 2016.

23. EEA Agency. *Europe's Onshore and Offshore Wind Energy Potential*; EEA Agency: København K, Denmark, 2016.

24. Towarowa Giełda Energii. Available online: https://tge.pl/ (accessed on 31 January 2018).

25. Box, G.E.P.; Jenkins, G.M.; Reinsel, G.C.; Ljung, G.M. *Time Series Analysis: Forecasting and Control*; Prentice Hall: Upper Saddle River, NJ, USA, 1994; ISBN 978-0133341867.

26. Fausette, L. *Fundamentals of Neural Networks-Architecture, Algorithms and Applications*; Florida Institute of Technology: Melbourne, FL, USA, 1994; ISBN 978-0133341867.

27. Madsen, K.; Nielsen, H.B.; Tingleff, O. *Methods for Non-Linear Least Squares Problems*, 2nd ed.; Technical University of Denmark: Lyngby, Denmark, April 2004.
28. Maciejowska, K.; Nitka, W.; Weron, T. *Enhancing Load, Wind and Solar Generation Forecasts in Day-Ahead Forecasting of Spot and Intraday Electricity Prices*; HSC Research Report; Wrocław University of Science and Technology: Wrocław, Poland, 2019.

Article

Deep Learning-Assisted Short-Term Load Forecasting for Sustainable Management of Energy in Microgrid

Arash Moradzadeh [1], Hamed Moayyed [2], Sahar Zakeri [1], Behnam Mohammadi-Ivatloo [1,*] and A. Pedro Aguiar [2]

[1]   Faculty of Electrical and Computer Engineering, University of Tabriz, Tabriz 5166/15731, Iran;
arash.moradzadeh@tabrizu.ac.ir (A.M.); s_zakeri@tabrizu.ac.ir (S.Z.)
[2]   Department of Electrical and Computer Engineering, University of Porto, 4000-008 Porto, Portugal;
hmoayyed@fe.up.pt (H.M.); pedro.aguiar@fe.up.pt (A.P.A.)
*   Correspondence: bmohammadi@tabrizu.ac.ir

Abstract: Nowadays, supplying demand load and maintaining sustainable energy are important issues that have created many challenges in power systems. In these types of problems, short-term load forecasting has been proposed as one of the management and energy supply modes in power systems. In this paper, after reviewing various load forecasting techniques, a deep learning method called bidirectional long short-term memory (Bi-LSTM) is presented for short-term load forecasting in a microgrid. By collecting relevant features available in the input data at the training stage, it is shown that the proposed procedure enjoys important properties, such as its great ability to process time series data. A microgrid in rural Sub-Saharan Africa, including household and commercial loads, was selected as the case study. The parameters affecting the formation of household and commercial load profiles are considered as input variables, and the total household and commercial load profiles of the microgrid are considered as the target. The Bi-LSTM network is trained by input variables to forecast the microgrid load on an hourly basis by recognizing the consumption pattern. Various performance evaluation indicators such as the correlation coefficient (R), mean squared error (MSE), and root mean squared error ($RMSE$) are utilized to analyze the forecast results. In addition, in a comparative approach, the performance of the proposed method is compared and evaluated with other methods used in similar studies. The results presented for the training phase show an accuracy of R = 99.81% for the Bi-LSTM network. The test and load forecasting stage are performed by the Bi-STLM network, with an accuracy of R = 99.34% and forecasting errors of MSE = 0.1042 and $RMSE$ = 0.3243. The results confirm the high performance of the proposed Bi-LSTM technique, with a high correlation coefficient when compared to other methods used for short-term load forecasting.

Keywords: energy management; microgrid; residential and commercial loads; short-term load forecasting; deep learning; bidirectional long short-term memory (Bi-LSTM)

Citation: Moradzadeh, A.; Moayyed, H.; Zakeri, S.; Mohammadi-Ivatloo, B.; Aguiar, A.P. Deep Learning-Assisted Short-Term Load Forecasting for Sustainable Management of Energy in Microgrid. *Inventions* **2021**, 6, 15. https://doi.org/10.3390/inventions6010015

Received: 3 January 2021
Accepted: 29 January 2021
Published: 3 February 2021

Publisher's Note: MDPI stays neutral with regard to jurisdictional claims in published maps and institutional affiliations.

1. Introduction

Nowadays, the significant increase in power consumption has led to the development of and fundamental changes in power grids. Power grids now consist of large-scale power plants based on nuclear power and fossil fuels. These power plants are considered to be the primary source of energy production [1,2]. In these forms of generation, unilateral energy transmission is conducted from centralized power plants (microgrids) to energy consumers. Due to several factors, including the possible lack of fossil fuels and the dangers of rising greenhouse gases, renewable energies like solar energy and wind energy are becoming much more popular as clean and novel energy sources [3,4]. To address this trend, many efforts were made to create a novel type of power system by combining communication and optimization theory to have optimal power management. A smart grid is a next-generation power grid dependent on various sources of energy, such as renewable energy. A smart grid aims to employ energy generation and consumption data by smart meters to manage energy efficiently [5,6].

Inventions **2021**, 6, 15. https://doi.org/10.3390/inventions6010015   https://www.mdpi.com/journal/inventions

Distributed generation (DG) was discussed as one of the specifications of renewable energy generation, which is done by different types of small, grid-connected, or distribution system-connected devices [7]. DGs are also commonly referred to as on-site generation, district energy, and distributed energy. Unlike traditional power plants, the generation and consumption of energy through renewable energy resources requires small-scale power plants which, in this case, will make the power system infrastructure more complex in general. In this situation, and to solve this problem, the microgrid is utilized as a building block to ensure the reliability and efficiency of the power system infrastructure [8].

In fact, a microgrid is a small-scale grid and a suitable solution for integrating variable and unpredictable renewable energy sources into distribution networks. Based on the structure and scale, a microgrid is very cost-effective in terms of infrastructure transfer [9]. Basically, the definition of a microgrid can be considered in such a way that, in order to form a self-sufficient energy system, it collects locally distributed generation sources, along with controllable loads and energy storage equipment. Short-term load forecasting is known as a principle function for the microgrid energy management system, particularly if different renewable energy resources are integrated with the microgrid. In addition, short-term forecasting can be considered an essential tool for microgrid operators to maintain continuous network performance and increase economic gains [10,11].

Short-term load forecasting is a function that has been performed in various ways so far. However, in the previous studies, one can mainly point out some conventional procedures that have been presented to forecast the load, which are as follows: persistence, statistical, physical, artificial neural network (ANN), machine learning, deep learning, and hybrid techniques [12–14]. In a valuable study [15], a variety of data-driven techniques were introduced and employed in a comparative approach to solve the necessary forecasting problems in the power grid. In [16], applications of ANN models were used to forecast the amount of wind and solar power in the microgrids. In a review paper [17], various types of ANN algorithms and issues with their application in the microgrid were reviewed. Short-term load forecasting in a microgrid was done in [14] by producing a novel hybrid technique based on support vector regression (SVR) and long short-term memory (LSTM) models. In [18], short-term load forecasting was performed using a hybrid technique of machine learning applications called seasonality-adjusted support vector machines (SSA-SVM). In [19], load forecasting in a microgrid was done using various deep learning algorithms, multilayer perceptron (MLP), and a support vector machine (SVM). In some other studies, combinations of machine learning models with optimization algorithms for load forecasting have been proposed. In [20], the microgrid load was forecasted using the hybridized model of an SVM and particle swarm optimization (PSO) in a short-term horizon. In another valuable study [21], a combined approach of a wavelet transform (WT) with a fruit fly optimization (FFO) algorithm for short-term load forecasting was proposed. In most cases, this type of hybrid model, due to the high dimension of input data, has more problems, such as overfitting, due to the time series characteristics of the relevant data not being able to identify the appropriate pattern of data.

Each of the reviewed studies attempted to forecast the load in a short-term time horizon using a variety of techniques. In some of them, the methods used were not commensurate with the available data and caused a decrease in the forecast accuracy due to factors such as overfitting and data missing in the training phase. In some others, the selected procedure was such that it suffered from time series feature modeling of the data, and this action reduced the accuracy of the forecasting. Most importantly, in some studies, the chosen method was not able to forecast the amount of load during peak hours, which caused problems in the network operator's scheduling.

It is noteworthy that deep learning techniques have been used as a powerful tool in recent studies. They have also performed well at preprocessing, processing, and extracting features from raw data and addressing the problems mentioned in short-term forecasting issues [22]. In previous studies, some deep learning techniques for processing and predicting time series data, such as convolutional neural networks (CNNs), deep autoencoders (DAEs), recurrent neural networks (RNNs), and deep belief neural networks (DBNNs), are presented. Each of these introduced techniques has some unique advantages and disadvantages. The DAE and DBNN techniques suffer at understanding long dependencies in time series samples related to anticipation time [23], while the CNN method with the least number of cells and memory can extract the basic features of the time series data. However, filter selection and the number of layers are issues that, if not properly selected, can cause problems such as overfitting in the training phase. The RNN-based techniques, such as LSTM and gated recurrent unit (GRU), usually perform well in time series data processing and can model complex time-dependent nonlinear parameters. The GRU networks mainly extract features that are not obtained by LSTM networks and are less complex than the LSTM. However, some studies have shown that the LSTM algorithms, due to forward training, have suffered from problems such as missing data and overfitting when recognizing patterns of large-volume data [24].

In this paper, in order to forecast the short-term load in the microgrid and solve the problems related to the reviewed methods, a deep learning techniques called bidirectional LSTM (Bi-LSTM) is proposed. Bi-LSTM is a time series-based technique that considers all data behavior in a time period. It should be noted that Bi-LSTM is proposed for the first time in this paper to forecast the short-term load in microgrids. Data affecting the network load have a long-term interconnected behavior and pattern. Accordingly, the bidirectional movement of the proposed method and the interconnected and related structure of its layers eliminates problems such as missing data and overfitting in the training phase. In comparing the Bi-LSTM technique with other models reviewed in the literature, some of the structural and inherent advantages of Bi-LSTM, such as learning the forward rule of data information as well as the backward rule of data information, indicate the strong performance of this technique.

In general, the contribution and practical tips of this paper can be highlighted as follows:

- Implementing a learning-based approach that, with its high skill, passes the training phase without problems such as missing data and overfitting;
- Forecasting microgrid load without considering meteorological data that are not available in remote areas;
- Modeling of microgrid load consumption for a short-term time horizon (one hour) based on different household and commercial consumption loads;
- Evaluating the performance of the Bi-LSTM technique in the training phase and the results of load forecasting with different performance evaluation indicators, as well as presenting a comparative approach to express the effectiveness of the suggested method.

In the next sections, the paper is organized as follows. The suggested method is described in Section 2. The case study is introduced in Section 3. The results of the short-term load forecasting are presented in Section 4. Finally, the conclusion of the paper is done in Section 5.

2. Bidirectional Long Short-Term Memory (Bi-LSTM)

In recent years, the application of deep learning has been significantly considered and used in various scientific and industrial fields. As such, deep learning techniques are used today in various applications in power and energy systems, such as fault detection [25,26], cyberattack detection [27], renewable power plant potential measurement [28], non-intrusive load monitoring [29,30], and load forecasting [14,31]. Deep learning has different techniques, each of which is skilled in specific applications due to its unique structure. In this paper, to solve short-term load forecasting in microgrids, one of the most powerful deep learning techniques, called Bi-LSTM, is proposed.

Bi-LSTM is a deep learning application used for classification, regression, pattern recognition, and feature extraction applications. One of the salient features of this technique is its excellent performance against time series data [32]. Bi-LSTM, as an extension of the traditional LSTM [33], is trained on the input sequence, with two LSTMs set up in reverse order (see Figure 1). The LSTM layer reduces the vanishing gradient problem and allows the use of deeper networks compared with recurrent neural networks (RNNs) [34,35].

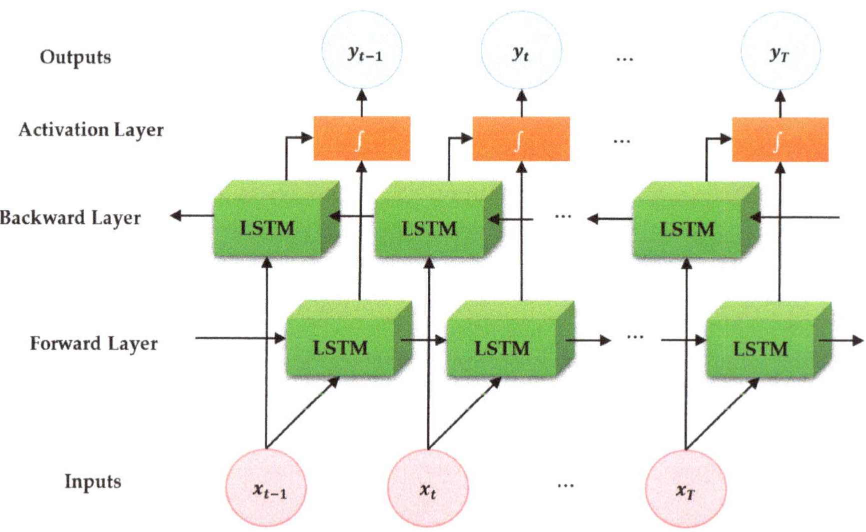

Figure 1. The bidirectional long short-term memory (Bi-LSTM) structure.

In the structure of the traditional RNN and the LSTM model, the propagation of information happens in a forward path, in which case the time t depends only on the information before the time t. In the Bi-LSTM network, unlike in traditional LSTM, flowing the information from the backward layer to the forward layer and upside down is performed by employing a hidden state [36]. Additionally, the advantage of Bi-LSTM over convolutional neural networks (CNNs) is its dependency on the sequence of inputs by taking the forward and backward paths into account. The Bi-LSTM model behaves the same with all inputs. The mathematical formulations of Bi-LSTM are presented in detail in [36].

3. Case Study

In this paper, a rural microgrid in Sub-Saharan Africa was selected as the case study. The specifications and data related to this microgrid, which was a freely available dataset, constituted the input variables and outputs of the dataset used in this paper [14]. The access and use of electrical energy for South African citizens is a human rights matter that is guaranteed by government policies. However, some problems, such as the lack of a sustainable electricity supply, plague many remote rural areas. Accordingly, this paper selected a microgrid in South Africa as the case study in order to provide solutions for energy management and sustainable energy supplies. The studied microgrid included household and commercial loads, which constituted the total load consumption of the microgrid. In the existing dataset, the household load was modeled based on factors such as the number of households (NoH) available and the percentage of high-income (HI), middle-income (MI), and low-income (LI) households. Factors such as water pumping (WP), grain milling (GM), and the amount of clinics, small shops (SS), schools, and street lighting (SL) also modeled the commercial load. The modeling was done to calculate the load on an hourly basis and in a one hour interval. As an example, Figure 2 shows three examples of 24 h load profiles under different conditions in this microgrid. Table 1 also introduces the prevailing conditions for the formation of load profiles, shown in Figure 2.

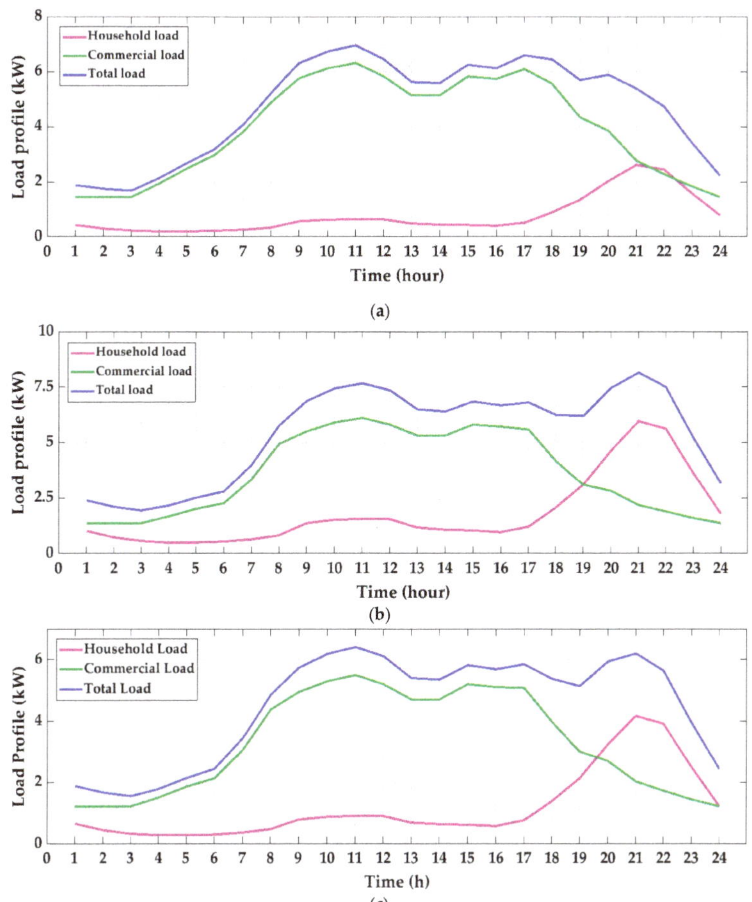

Figure 2. Samples of 24 h microgrid load profiles for (**a**) a low-population microgrid with the same income percentage for the households, (**b**) a populous microgrid with high-income households, and (**c**) a populous microgrid with low-income households.

Table 1. Input parameters for calculating each instance of the microgrid load profiles shown in Figure 2.

Input Variable	Figure 2a	Figure 2b	Figure 2c
NoH	50	100	100
HI households	33%	45%	20%
MI households	33%	30%	30%
LI households	33%	25%	50%
Number of WP	5	3	3
Number of GM	1	1	1
Number of SS	5	3	3
Number of Schools	1	3	2
Number of Clinics	1	3	3
Number of SL	10	15	12

4. Simulation Results

Short-term load forecasting in a microgrid using the Bi-LSTM method required a dataset containing effective hourly data on the microgrid load and the related load profiles. Hourly parameters related to the characteristics of households and equipment and commercial places were considered as input variables, and the hourly load profile resulting from these characteristics was selected as the output variable. The existing dataset contained 240 samples of 24 h load profiles which, by considering the data associated with each hour as an input sample, would eventually form 5760×11 matrix for the Bi-LSTM input dataset. The designed Bi-LSTM network was trained by 70% of the data. Then, in the test phase, using the rest of the data, it forecasted the microgrid load over the one-hour intervals.

After the training and test stages, the results of each learning-based network should be evaluated using performance appraisal indicators. This expresses the accuracy of each network at each step and clearly indicates how close the forecasted or estimated values are to the actual values. In this paper, the Bi-LSTM network performance was evaluated by indicators such as the correlation coefficient (*R*), mean squared error (*MSE*), and root mean squared error (*RMSE*). The R-index showed a kind of correlation between the forecasted values and the real values, and the maximum values of *R* indicated the high accuracy of the network. The *MSE* and *RMSE* indicators showed the prediction error that were calculated for these indices for each sample and, finally, a mean value was calculated for the network performance in the whole dataset. The proximity of the *MSE* and *RMSE* indicator values to zero indicated the accuracy of the network performance [37]. The mathematical formulation for calculating each of the indicators used in this paper is as follows [38]:

$$R = \frac{\sum_{i=1}^{N}(x_i - \overline{x})(y_i - \overline{y})}{\sqrt{\sum_{i=1}^{N}(x_i - \overline{x})^2 \sum_{i=1}^{N}(y_i - \overline{y})^2}} \tag{1}$$

$$MSE = \frac{\sum_{i=1}^{N}(x_i - y_i)^2}{N} \tag{2}$$

$$RMSE = \sqrt{\frac{\sum_{i=1}^{N}(x_i - y_i)^2}{N}} \tag{3}$$

where x_i and y_i represent the actual values and forecasted values, respectively, and $\overline{x}$ and $\overline{y}$ are the means of the actual values and forecasted values, respectively.

Figure 3 shows the performance of the Bi-LSTM network in the training phase using the R-index and in regression form.

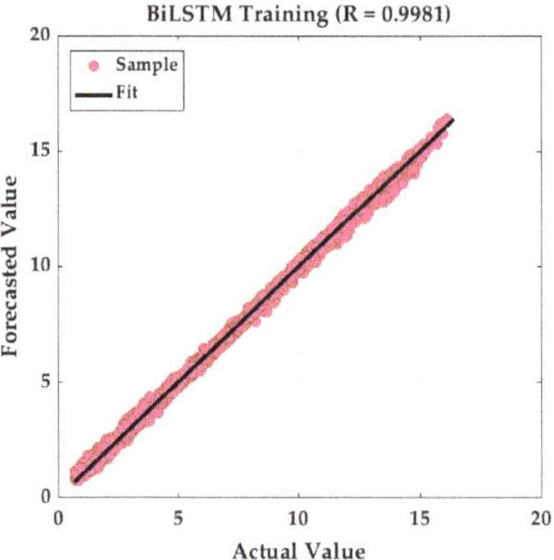

Figure 3. Bi-LSTM performance in the training phase in the form of regression.

As shown in Figure 3, the Bi-LSTM network was able to pass the training stage with good performance. Figure 4 shows the amount of network error in the training phase. It can be seen that the training error rate was very small, and these results indicated good network training. When trained with high accuracy, the network would be able to ideally identify test data with an estimated model and forecast their values well.

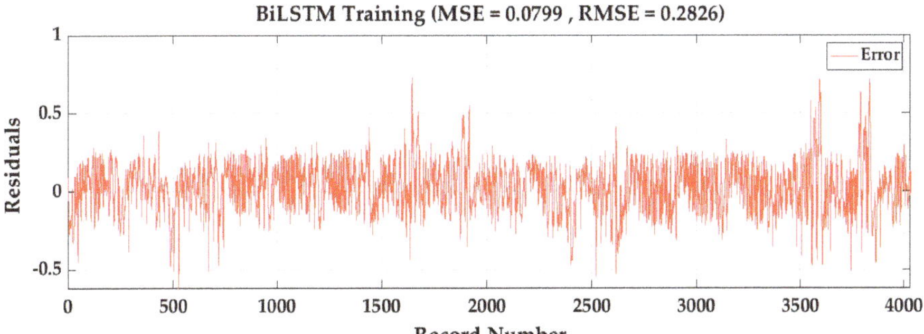

Figure 4. Bi−LSTM training errors in the forms of the mean squared error (*MSE*) and root mean squared error (*RMSE*).

After training, the test data, which was 30% of the input data set, was used for prediction as input to the trained network. At this stage, the trained Bi-LSTM estimated the load profile for each sample based on the data behavior pattern in the training stage. Figure 5a shows the load forecasting results by Bi-LSTM in the test stage. In order to clearly observe the performance of the proposed method in forecasting the microgrid load, Figure 5b shows 100 samples related to the test data in zoom mode, which is presented in Figure 5a.

The forecasted load was, in fact, the sum of the household and commercial loads related to the microgrid at each hour. The high correlation of the forecasted values with the actual values of the load profile confirmed the good performance of the trained network. It can be seen that the network accuracy coefficient was an acceptable value (*R* = 0.9934), and the amount of microgrid consumption could be estimated at any time. Figures 6 and 7 also show the network prediction error in the forms of the *MSE*, *RMSE*, and a histogram, respectively.

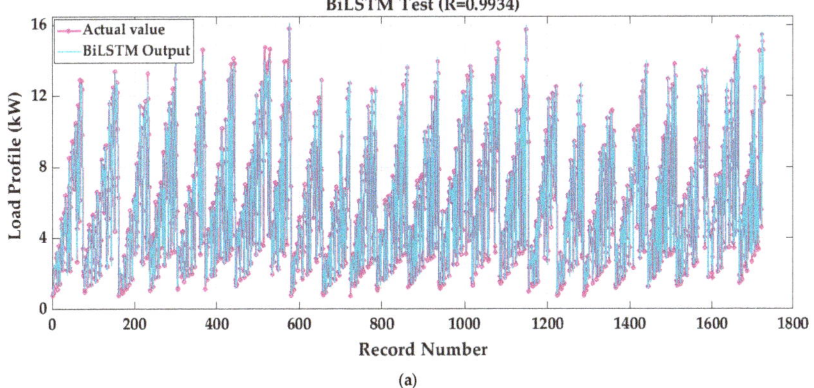

Figure 5. *Cont.*

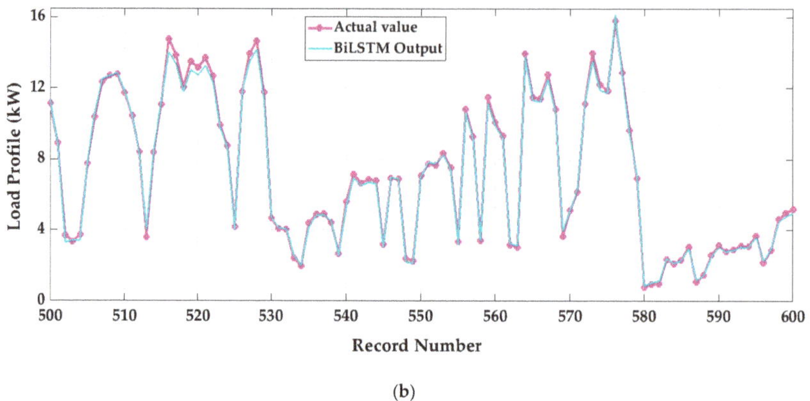

(**b**)

Figure 5. Load forecasting results by Bi-LSTM in the test stage. (**a**) All test data. (**b**) Zoomed image of the forecasting of 100 test data samples.

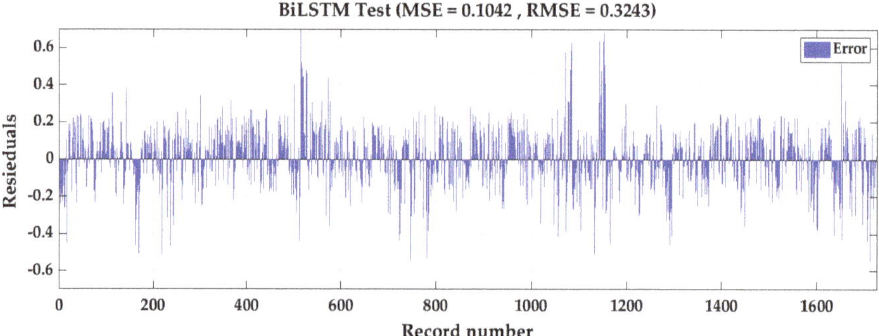

Figure 6. Bi-LSTM test errors in the forms of the *MSE* and *RMSE*.

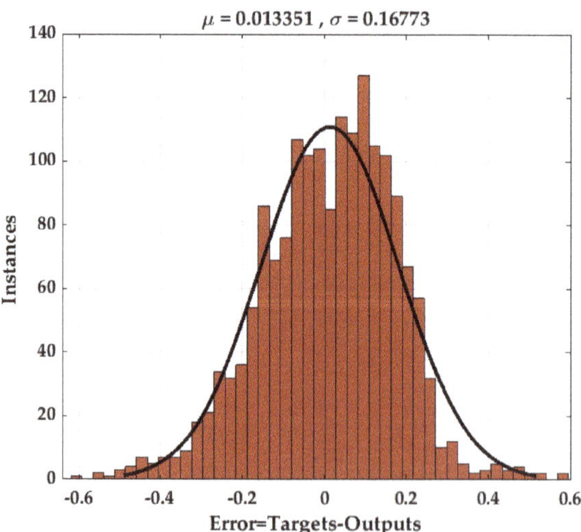

Figure 7. Bi−LSTM test error in the form of a histogram.

According to Figure 6, the amount of network prediction error in each sample can be seen. The error values obtained in this figure (e.g., $MSE = 0.1042$ and $RMSE = 0.3243$) were the averages of the prediction errors in all samples. Figure 7 also shows the minimum and maximum network errors in forecasting the value of each sample. One can see that, in the worst-case scenario, the largest error by the Bi-LSTM was in forecasting the microgrid load between the numbers -0.6 and 0.6.

After presenting the results obtained with the proposed Bi-LSTM network and evaluating its performance with other statistical indicators, the comparison with other similar works to confirm its effectiveness proceeded. To this end, the results of similar works presented in recent years were compared with the results obtained by Bi-LSTM via performance evaluation indicators. Table 2 shows the results of this comparison.

Table 2. Evaluation of the performance of the proposed method in comparison with other solutions presented in similar studies.

Table	R	MSE	$RMSE$
Bi-LSTM (Proposed)	0.9934	0.1042	0.3243
SVR [14]	0.9770	0.5983	0.7735
LSTM [14]	0.9809	0.5133	0.7164
SVR-LSTM [14]	0.9901	0.1316	0.3627
SVM [19]	-	-	6.1910
MLP [19]	-	-	5.6540
CNN-GRU [39]	-	-	0.1617
CNN-LSTM [39]	-	-	0.2265

In the evaluation performed in Table 2, the performance of different machine learning and deep learning methods were compared with each other. As can be seen, deep learning methods offer better performance than more conventional machine learning methods. The reason for this superiority is proper training and extracting the appropriate pattern from the input data. Among the compared methods, the Bi-STM network proposed in this paper was able to surpass other methods with better performance. In using data mining techniques, choosing the right method with the available data is one of the most important issues. The proposed Bi-LSTM procedure can be selected as a tool to perform other time series-based data predictions in power and energy systems.

5. Conclusions

Short-term load forecasting in power and energy systems is a key technique to improve the supplying of a demand load and other energy management planning. The purpose of this paper was to forecast the short-term load in a rural microgrid in Sub-Saharan Africa in order to supply the demand load and access to sustainable energy. To this end, one of the deep learning algorithms, called bidirectional long short-term memory (Bi-LSTM), was proposed. Unlike other deep learning techniques, the Bi-LSTM method, due to its unique structure and bidirectional training routine, offers a strong ability to process large-volume and time series data. In addition, avoiding the problems of missing data and overfitting during the training phase can be mentioned as other benefits of Bi-LSTM. Data related to household and commercial loads were collected in the studied system as a Bi-LSTM input dataset. The Bi-LSTM network was trained with the input data and then forecasted the microgrid load in one-hour intervals. The results of the forecast were analyzed by various performance evaluation indicators such as the correlation coefficient (R), mean squared error (MSE), and root mean squared error ($RMSE$). The trained LSTM network was able to forecast the microgrid load in the short-term time horizon with an accuracy of $R = 80\%$ and the lowest error values (in this case, $MSE = 0.1042$ and $RMSE = 0.3243$). Then, in a comparative approach, the results of the Bi-LSTM network were evaluated with other algorithms used in similar works. The results of the evaluations emphasized the effectiveness of the suggested procedure in the short-term load forecasting of microgrids compared with other solutions.

One future work direction is the application, implementation, and evaluation of the proposed methodology for the load forecasting of other microgrids in the presence of renewable energy resources.

Author Contributions: A.M.: writing—original draft, software, methodology, and data curation; H.M.: formal analysis, investigation and validation, and conceptualization; S.Z.: methodology, investigation and validation, and resources; B.M.-I. and A.P.A.: formal analysis, investigation and validation, conceptualization, and writing—review and editing. All authors have read and agreed to the published version of the manuscript.

Funding: This work was supported in part by the Portuguese Foundation for Science and Technology through project IMPROVE (POCI-01-0145-FEDER-031823), funded by the FEDER Funds through the COMPETE2020—POCI, and in part by the National Funds (PIDDAC).

Conflicts of Interest: The authors declare no conflict of interest.

References

1. Ihsan, A.; Brear, M.J.; Jeppesen, M. Impact of operating uncertainty on the performance of distributed, hybrid, renewable power plants. *Appl. Energy* **2021**, *282*, 116256. [CrossRef]
2. Caliano, M.; Bianco, N.; Graditi, G.; Mongibello, L. Design optimization and sensitivity analysis of a biomass-fired combined cooling, heating and power system with thermal energy storage systems. *Energy Convers. Manag.* **2017**, *149*, 631–645. [CrossRef]
3. Akinyele, D.; Olabode, E.; Amole, A. Review of fuel cell technologies and applications for sustainable microgrid systems. *Inventions* **2020**, *5*, 42. [CrossRef]
4. Sadeghian, O.; Moradzadeh, A.; Mohammadi-Ivatloo, B.; Abapour, M.; Marquez, F.P.G. Generation units maintenance in combined heat and power integrated systems using the mixed integer quadratic programming approach. *Energies* **2020**, *13*, 2840. [CrossRef]
5. Khalid, H.; Shobole, A. Existing Developments in Adaptive Smart Grid Protection: A Review. *Electric Power Syst. Res.* **2021**, *191*, 106901. [CrossRef]
6. Yan, B.; Di Somma, M.; Graditi, G.; Luh, P.B. Markovian-based stochastic operation optimization of multiple distributed energy systems with renewables in a local energy community. *Electric Power Syst. Res.* **2020**, *186*, 106364. [CrossRef]
7. Ghorbani, S.; Unland, R.; Shokouhandeh, H.; Kowalczyk, R. An innovative stochastic multi-agent-based energy management approach for microgrids considering uncertainties. *Inventions* **2019**, *4*, 37. [CrossRef]
8. Fayek, H.H.; Mohammadi-Ivatloo, B. Tidal Supplementary Control Schemes-Based Load Frequency Regulation of a Fully Sustainable Marine Microgrid. *Inventions* **2020**, *5*, 53. [CrossRef]
9. Khan, M.; Khan, M.; Jiang, H.; Hashmi, K.; Shahid, M. An Improved Control Strategy for Three-Phase Power Inverters in Islanded AC Microgrids. *Inventions* **2018**, *3*, 47. [CrossRef]
10. Lee, E.-K.; Shi, W.; Gadh, R.; Kim, W. Design and Implementation of a Microgrid Energy Management System. *Sustainability* **2016**, *8*, 1143. [CrossRef]
11. Tayab, U.B.; Zia, A.; Yang, F.; Lu, J.; Kashif, M. Short-term load forecasting for microgrid energy management system using hybrid HHO-FNN model with best-basis stationary wavelet packet transform. *Energy* **2020**, *203*, 117857. [CrossRef]
12. Zhang, Y.; Le, J.; Liao, X.; Zheng, F.; Li, Y. A novel combination forecasting model for wind power integrating least square support vector machine, deep belief network, singular spectrum analysis and locality-sensitive hashing. *Energy* **2019**, *168*, 558–572. [CrossRef]
13. Liang, Y.; Niu, D.; Hong, W.C. Short term load forecasting based on feature extraction and improved general regression neural network model. *Energy* **2019**, *166*, 653–663. [CrossRef]
14. Moradzadeh, A.; Zakeri, S.; Shoaran, M.; Mohammadi-Ivatloo, B.; Mohammadi, F. Short-term load forecasting of microgrid via hybrid support vector regression and long short-term memory algorithms. *Sustainability* **2020**, *12*, 7076. [CrossRef]
15. Ferlito, S.; Adinolfi, G.; Graditi, G. Comparative analysis of data-driven methods online and offline trained to the forecasting of grid-connected photovoltaic plant production. *Appl. Energy* **2017**, *205*, 116–129. [CrossRef]
16. Heydari, A.; Astiaso Garcia, D.; Keynia, F.; Bisegna, F.; De Santoli, L. A novel composite neural network based method for wind and solar power forecasting in microgrids. *Appl. Energy* **2019**, *251*, 113353. [CrossRef]
17. Lopez-Garcia, T.B.; Coronado-Mendoza, A.; Domínguez-Navarro, J.A. Artificial neural networks in microgrids: A review. *Eng. Appl. Artif. Intell.* **2020**, *95*, 103894. [CrossRef]
18. Che, J.; Wang, J. Short-term load forecasting using a kernel-based support vector regression combination model. *Appl. Energy* **2014**, *132*, 602–609. [CrossRef]
19. Wen, L.; Zhou, K.; Yang, S.; Lu, X. Optimal load dispatch of community microgrid with deep learning based solar power and load forecasting. *Energy* **2019**, *171*, 1053–1065. [CrossRef]

20. Selakov, A.; Cvijetinović, D.; Milović, L.; Mellon, S.; Bekut, D. Hybrid PSO-SVM method for short-term load forecasting during periods with significant temperature variations in city of Burbank. *Appl. Soft Comput. J.* **2014**, *16*, 80–88. [CrossRef]
21. Sun, W.; Ye, M. Short-Term Load Forecasting Based on Wavelet Transform and Least Squares Support Vector Machine Optimized by Fruit Fly Optimization Algorithm. *J. Electrical Comput. Eng.* **2015**, *2015*, 862185. [CrossRef]
22. Teng, W.; Cheng, H.; Ding, X.; Liu, Y.; Ma, Z.; Mu, H. DNN-based approach for fault detection in a direct drive wind turbine. *IET Renew. Power Gener.* **2018**, *12*, 1164–1171. [CrossRef]
23. Zhang, S.; Zhang, S.; Wang, B.; Habetler, T.G. Deep Learning Algorithms for Bearing Fault Diagnostics—A Comprehensive Review. *IEEE Access* **2020**, *8*, 29857–29881. [CrossRef]
24. Schmidhuber, J. Deep Learning in neural networks: An overview. *Neural Netw.* **2015**, *61*, 85–117. [PubMed]
25. Moradzadeh, A.; Pourhossein, K. Short Circuit Location in Transformer Winding Using Deep Learning of Its Frequency Responses. In Proceedings of the 2019 International Aegean Conference on Electrical Machines and Power Electronics, ACEMP 2019 and 2019 International Conference on Optimization of Electrical and Electronic Equipment, OPTIM 2019, Istanbul, Turkey, 27–29 August 2019; IEEE: New York, NY, USA, 2019; pp. 268–273.
26. Teimourzadeh, H.; Moradzadeh, A.; Shoaran, M.; Mohammadi-Ivatloo, B.; Razzaghi, R. High Impedance Single-Phase Faults Diagnosis in Transmission Lines via Deep Reinforcement Learning of Transfer Functions. *IEEE Access* **2021**, *9*, 15796–15809. [CrossRef]
27. Yu, J.J.Q.; Hou, Y.; Li, V.O.K. Online False Data Injection Attack Detection With Wavelet Transform and Deep Neural Networks. *IEEE Trans. Ind. Inform.* **2018**, *14*, 3271–3280. [CrossRef]
28. Wang, H.; Lei, Z.; Zhang, X.; Zhou, B.; Peng, J. A review of deep learning for renewable energy forecasting. *Energy Convers. Manag.* **2019**, *198*, 111799. [CrossRef]
29. Kong, W.; Dong, Z.Y.; Wang, B.; Zhao, J.; Huang, J. A practical solution for non-intrusive type II load monitoring based on deep learning and post-processing. *IEEE Trans. Smart Grid* **2020**, *11*, 148–160. [CrossRef]
30. Moradzadeh, A.; Mohammadi-Ivatloo, B.; Abapour, M.; Anvari-Moghaddam, A.; Gholami Farkoush, S.; Rhee, S.B. A practical solution based on convolutional neural network for non-intrusive load monitoring. *J. Ambient Intell. Humaniz. Comput.* **2021**. [CrossRef]
31. Ebrahim, A.F.; Mohammed, O.A. Pre-processing of energy demand disaggregation based data mining techniques for household load demand forecasting. *Inventions* **2018**, *3*, 45. [CrossRef]
32. Shrestha, A.; Mahmood, A. Review of deep learning algorithms and architectures. *IEEE Access* **2019**, *7*, 53040–53065. [CrossRef]
33. Hochreiter, S.; Schmidhuber, J. Long Short-Term Memory. *Neural Comput.* **1997**, *9*, 1735–1780. [CrossRef] [PubMed]
34. Zakeri, S.; Geravanchizadeh, M. Supervised binaural source separation using auditory attention detection in realistic scenarios. *Appl. Acoust.* **2021**, *175*, 107826. [CrossRef]
35. Bengio, Y.; Simard, P.; Frasconi, P. Learning long-term dependencies with gradient descent is difficult. *IEEE Trans. Neural Netw.* **1994**, *5*, 157–166. [CrossRef]
36. Xie, W.; Wang, J.; Xing, C.; Guo, S.; Guo, M.; Zhu, L. Variational Autoencoder Bidirectional Long and Short-term Memory Neural Network Soft-sensor Model Based on Batch Training Strategy. *IEEE Trans. Ind. Inform.* **2020**. [CrossRef]
37. Moradzadeh, A.; Mansour-Saatloo, A.; Mohammadi-Ivatloo, B.; Anvari-Moghaddam, A. Performance evaluation of two machine learning techniques in heating and cooling loads forecasting of residential buildings. *Appl. Sci.* **2020**, *10*, 3829. [CrossRef]
38. Mansour-Saatloo, A.; Moradzadeh, A.; Mohammadi-Ivatloo, B.; Ahmadian, A.; Elkamel, A. Machine learning based PEVs load extraction and analysis. *Electronics* **2020**, *9*, 1150. [CrossRef]
39. Afrasiabi, M.; Mohammadi, M.; Rastegar, M.; Kargarian, A. Multi-agent microgrid energy management based on deep learning forecaster. *Energy* **2019**, *186*, 115873. [CrossRef]

Article

Techno-Economic Comparative Analysis of Renewable Energy Systems: Case Study in Zimbabwe

Loiy Al-Ghussain [1,*], Remember Samu [2], Onur Taylan [3,4] and Murat Fahrioglu [4,5]

[1]   Mechanical Engineering Department, University of Kentucky, Lexington, KY 40506, USA
[2]   Discipline of Engineering and Energy, College of Science, Health, Engineering and Education, Murdoch University, South Street, Murdoch, WA 6150, Australia; 33635333@student.murdoch.edu.au
[3]   Mechanical Engineering Program, Middle East Technical University Northern Cyprus Campus, Guzelyurt via Mersin 10, 99738 Kalkanli, Turkey; ontaylan@metu.edu.tr
[4]   Center for Solar Energy Research and Applications (GÜNAM), Middle East Technical University, 06800 Ankara, Turkey; fmurat@metu.edu.tr
[5]   Electrical and Electronics Engineering Program, Middle East Technical University Northern Cyprus Campus, Guzelyurt via Mersin 10, 99738 Kalkanli, Turkey
*   Correspondence: loiy.al-ghussain@uky.edu

Received: 12 June 2020; Accepted: 3 July 2020; Published: 6 July 2020

Abstract: Fluctuations in fossil fuel prices significantly affect the economies of countries, especially oil-importing countries, hence these countries are thoroughly investigating the increase in the utilization of renewable energy resources as it is abundant and locally available in all the countries despite challenges. Renewable energy systems (RES) such as solar and wind systems offer suitable alternatives for fossil fuels and could ensure the energy security of countries in a feasible way. Zimbabwe is one of the African countries that import a significant portion of its energy needs which endanger the energy security of the country. Several studies in the literature discussed the feasibility of different standalone and hybrid RES either with or without energy storage systems to either maximize the technical feasibility or the economic feasibility; however, none of the studies considered maximizing both feasibilities at the same time. Therefore, we present a techno-economic comparison of standalone wind and solar photovoltaic (PV) in addition to hybrid PV/wind systems based on maximizing the RES fraction with levelized cost of electricity (LCOE) being less than or equal to the local grid tariff where Gwanda, Zimbabwe, is the case study. The methodology suggested in this study could increase the utilization of renewable energy resources feasibly and at the same time increase the energy security of the country by decreasing dependency on imported energy. The results indicate that the PV/wind hybrid system does not only have the best economic benefits represented by the net present value (NPV) and the payback period (PBP), but also the best technical performance; where the maximum feasible size of the hybrid system-2 MW wind and 1 MW PV-has RES fraction of 65.07%, LCOE of 0.1 USD/kWh, PBP of 3.94 years, internal rate of return of 14.04% and NPV of 3.06×10^6 USD. Having similar systems for different cities in Zimbabwe will decrease the energy bill significantly and contribute toward the energy security of the country.

Keywords: hybrid systems; photovoltaic; wind energy; energy economics; RES investments; Zimbabwe; Africa and energy security

---

1. Introduction

Environmental protection, energy resources conservation, and sustainable energy development are the core challenges that the world is facing nowadays [1] especially in hard times when crises have prevailed [2,3]. Moreover, given that the demand for electricity or rather energy is continuously increasing, being able to meet this demand in an environmentally friendly manner is of importance.

Enterprises are constantly searching for upgrades in their procedures for better use [4–6]. Moreover, the issue is taken care of by utilizing clean power [7,8], usage of waste power [9] and implementing various arrangements [10,11] where assets and condition are preserved. Henceforth, adjustments on power systems to expand force and productivity are required [12].

Zimbabwe faces a deficit in its energy production. Around 35% of the energy used in Zimbabwe is imported from the Democratic Republic of Congo (DRC), South Africa and Mozambique [13]. Renewable energy can be an environmentally friendly and sustainable solution to help curb this energy deficit in Zimbabwe. Basic knowledge of the geographical location, economic situation, as well as energy demand, is required to design and develop suitable systems. Zimbabwe lies in a sunny belt, with approximately 4000 h of solar radiation per year and 5.5 kWh/m^2/day of solar radiation on average [14,15]. Renewable energy has not been harnessed on a large scale in Zimbabwe. Small-scale solar photovoltaic (PV) panels for lighting and irrigation purposes are the only uses of solar energy in Zimbabwe [13]. Small windmills still exist on some farms, in which they were once used to pump water even though a study by the ZERO Regional Environment Organization concluded that there is a potential of generating power from wind resources in Zimbabwe at a hub height of 80 m [13].

Renewable energy systems comprising individual resources and/or more hybrid power systems have been a topic of recent research worldwide [12,16–23]. The early research papers on hybrid power systems were published in the mid-1980s [24]. With the advances in research, in the early 1990s wind resources could now be harnessed on a commercial scale. The problem of intermittency, grid stability and reliability led to more expansion of literature on renewables as well as hybrid systems [25].

In literature, noteworthy efforts on feasibility studies on solar PV, wind and hybrid PV-wind power systems have been made both grid-connected or stand-alone systems [17,18,26–33]. Research on hybrid systems has been undertaken but the goals of each research project differ. For two different locations in Jordan [34,35] used MATLAB and HOMER software to model economically feasible hybrid solar PV-wind power systems. Their main goal was to determine the cost of energy (COE), the net present value (NPV) and the renewable energy source (RES) fraction.

Benlouba and Bourouis [36] studied the economic and technical feasibility of different off-grid PV-wind hybrid configurations with diesel generators for a village in Algeria. Moreover, Ashok [37] found the configuration of a PV-wind hybrid system that has the lowest cost of electricity for a village in India using a Quasi-Newtonian method. He found out that a PV/wind/diesel/micro hydro hybrid system would provide electricity for a whole day at a cost of 0.14 USD/kWh. Furthermore, Samu et al. [38] concluded that the LCOE of their hybrid PV-wind power system was greater than the grid tariff solely because they used a wind turbine prototype with high cut-in speed and the competing generation sources, coal and hydro are cheaper as well.

The first research on the analysis of the solar potential of the whole of Zimbabwe was undertaken by [13]. They concluded that solar PV generation is both economically and environmentally feasible in the whole of Zimbabwe. Further studies of solar PV resources in some parts of Africa were undertaken by [15,39–44]. In summary, solar is economically feasible in various parts of Africa. Studies on possible solar home systems to alleviate the current power challenges in Zimbabwe have been performed by Chahuruva and Dei [45]. In this study, the authors performed an experiment at Ashikaga Institute in Japan to obtain results which they believed could be transferable to Zimbabwe. Results of this study could be questionable since the geographical and weather conditions in Japan are different from those in Zimbabwe. Additionally, this analysis was performed at irradiation of 8.25 kWh/m^2/day against Zimbabwe's average irradiation of 5.72 kWh/m^2/day. Additionally, a study on the potential of Concentrated Solar Power (CSP) in Zimbabwe was performed and it was concluded that 71.4 GW can be generated from CSP [46]. A wind map of Africa was developed by [47] and concluded that on average Africa has an onshore wind potential at a height of 80 m through Geographic Information System (GIS) analysis. Another wind resource mapping exercise was undertaken by [48] for Kenya and the southern part of Africa. Research on the techno-economic potential of wind energy has been carried out for a remote area of Sahel Zone in Cameroon by [49,50] did an assessment for another location in Cameroon as well and determined that the total energy produced annually could reduce carbon dioxide emissions by 1200 tons per year. Their calculated results and those of the Wind Atlas Analysis and Application Program (WAsP) were in good correlation [50].

The goals and objectives of the Zimbabwe National Renewable Energy Policy (NREP) included the installation of 1100 MW of renewables by 2025 and 2100 MW of renewables by 2030. Renewables in this context referred to grid-connected solar PV, grid-connected wind, small hydro and bagasse. The policy reported the provision of tax and sale of power to third-party incentives by the government and also reduced license fees for renewable energy projects. However, the policy does not outline any possible feed-in tariffs for renewable energy resources which might still make investments unattractive. Additionally, the policy still does not mention any development of stand-alone microgrids to electrify remote areas in which the rural electrification rate is only 13% [51]. The NREP also reports that the Zimbabwean government will introduce mechanisms for funding renewable energy systems as well as implementing a renewable energy technologies program that encourages Independent Power Producers (IPPs) to invest in renewable energy projects in Zimbabwe. Additionally, a fund is to be established by the Ministry of Energy to promote solar energy to address the electricity crisis.

All of the papers reviewed for this study did not compare the economic and technical feasibilities of wind, PV and PV-wind hybrid system at the same location in order to determine the most suitable system and the most profitable one. Moreover, none of the studies in the literature considered maximizing both the technical and economic feasibilities of different standalone and hybrid PV/wind systems in Zimbabwe at the same time. Therefore, this study aims to perform a techno-economic comparative study of an on-grid wind, PV and PV/wind hybrid system to determine the best RES configuration to be installed where Gwanda, Zimbabwe, is the case study knowing that this methodology can be used to determine the best configuration to install in any region in the world. The optimal RES configurations were found based on maximizing the RES fraction with levelized cost of electricity being less than or equal to the local grid tariff; such a constraint will ensure the maximum environmental benefits of the systems, increase the energy security of the country and at the same time ensures the economic feasibility of the alternative energy systems. Such methodology will increase the utilization of the local and abundant renewable energy resources in oil-importing countries such as Zimbabwe which would count toward significant cuts in the imported energy bill. Moreover, the proposed methodology is in accordance with the NREP goals and provides a pathway to achieve these goals in Zimbabwe.

2. Theory and Methodology

2.1. Photovoltaic (PV) Energy Model

Energy production from the PV plant is affected by the ambient conditions; where the effect of the ambient temperature on the module efficiency is the only ambient condition considered in this study. The efficiency of the PV module can be estimated using Equation (1) [52].

$$\eta_{PV} = \eta_{PV,R} \times \left[1 - \beta_R \times \left(T_a + (T_{NOC} - T_{R,NOC}) \times \frac{I_T}{I_R} - T_{R,STC} \right) \right] \tag{1}$$

where η_{PV} is the photovoltaic module efficiency, $\eta_{PV,R}$ is the module reference efficiency, β_R is the temperature coefficient (1/°C), T_a is the ambient temperature (°C) which was obtained for Gwanda city using Meteonorm v7.1 software which generates a typical meteorological year (TMY) [52], T_{NOC} is the nominal operating cell temperature (°C), $T_{R,NOC}$ is the reference module temperature at nominal conditions (°C), I_T the total irradiation on a tilted surface (Wh m^{-2}), I_R is the reference irradiation at nominal conditions (Wh m^{-2}) and $T_{R,STC}$ is the reference module temperature at standard test conditions (°C). In this study, PV modules from Canadian Solar company (Guelph, Ontario, Canada) type CS6K-285M were used [53].

After estimating the global insolation on the photovoltaic module and, with the estimation of the PV module efficiency, the hourly energy generation from the photovoltaic plant can be found.

Using the methodology in Duffie and Beckman [54], the global insolation was estimated which was not repeated in this study for brevity. The hourly energy generated, E_P, can be estimated as,

$$E_P = \eta_{PV} \times I_T \times A_m \times N_m \tag{2}$$

where A_m is the single module area [m^2] and N_m is the number of PV modules.

2.2. Wind Energy Model

The wind shear coefficient (α) can represent the factors that affect the wind speed at hub height like the speed at ground level, the hub height, the time (hour, day, season), the nature of the terrain and the ambient temperature. α can be taken as $\frac{1}{7}$ if the specific data of the site is not available [55]. At hub height, the wind speed (u_Z) can be extrapolated as,

$$u_Z = u_g \times \left(\frac{Z}{Z_g}\right)^{\alpha} \tag{3}$$

where u_g is the wind speed at ground level (m/s), Z is the height of the hub (m) and Z_g is the ground level height (m) at which speed is measured and it is equal 10 m. The hourly wind speeds at ground level for Gwanda city were generated using Meteonorm v7.1 software. Figure 1 shows the average hourly wind speeds at 10 m in addition to the average daily global insolation on a horizontal surface in Gwanda.

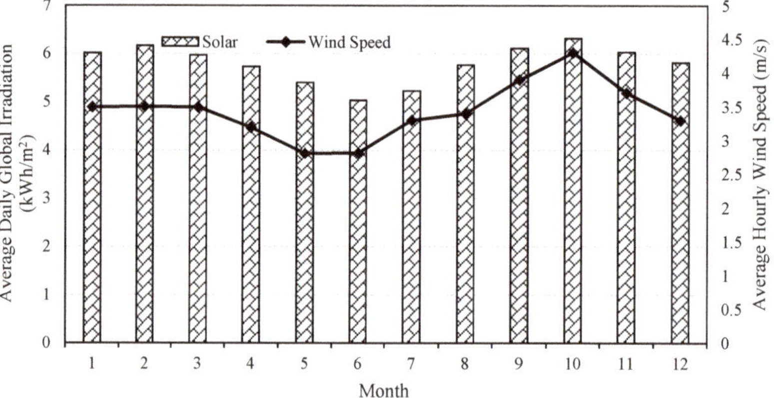

Figure 1. The average hourly wind speeds at ground level in addition to the average daily solar radiation on a horizontal surface throughout the year in Gwanda.

It is assumed that the energy generated from each wind turbine-in the case of multiple wind turbines-is the same; moreover, it is assumed that the energy production does not change during the hour. E_w which is the hourly energy production can be estimated using Equation (4).

$$E_w = \begin{cases} 0 \, , \ u_Z \langle u_C \ or \ u_Z \rangle u_F \\ N \times P_R \times \dfrac{(u_C)^K - (u_Z)^K}{(u_C)^K - (u_R)^K} \, , \ u_c \leq u_Z \leq u_R \\ N \times P_R \, , u_R < u_Z \leq u_F \end{cases} \tag{4}$$

where P_R is the wind turbine rated power (kW), N is the number of turbines, u_C is the wind turbine cut-in speed (m/s), u_F is the furling speed (m/s), K is the shape parameter of Weibull distribution and it can be calculated based on Justus' theory using Equation (5) and u_R is the rated wind speed (m/s).

2 MWwind turbine (G114-2.0) from Siemens Gamesa Renewable Energy company (Vizcaya, Spain) [56] was used in this study.

$$K = (\sigma/\overline{u})^{-1.086}, \; 1 \leq K \leq 10. \tag{5}$$

where $\overline{u}$ is the mean wind speed at hub level (m/s), and σ is the standard deviation of the wind speeds sample (m/s).

2.3. Performance Assessment of the RES

The harmony between the energy generation from the RESs and the demand can be inspected using the RES Fraction (F_R) which represent the percentage of demand met by the RES. F_R can be calculated as,

$$F_R = \frac{D_R}{D} \tag{6}$$

where D_R is the hourly demand met by the RES (kWh) and D is the hourly demand of Gwanda city (kWh) where the daily demand of Gwanda was obtained from [38] while the hourly load profile for Gwanda city was approximated to be like the profile of a city in Zambia [57].

Moreover, in order to inspect the autonomy of the system the demand supply fraction (DSF) which is the number of hours in which the demand is totally met by the RES in a year over the total numbers of hours in a year and it can be calculated using Equation (7).

$$DSF = \frac{H}{8760} \tag{7}$$

where DSF is the demand supply fraction (%); H is the number of hours in which the demand is totally met by the RES in a year.

2.4. The Economic Assessment of the Renewable Energy Systems (RES)

The levelized cost of electricity (LCOE) is used to assess the economic feasibility of the RESs. LCOE of the energy systems can be found using Equation (8) where the effect of the mismatching between the demand and energy production is incorporated by using the demand met by the hybrid system instead of the energy generated.

$$LCOE = \frac{C_i + \sum_{t=1}^{L} \frac{M_t}{(1+r)^t}}{\sum_{t=1}^{L} \frac{D_R}{(1+r)^t}} \tag{8}$$

where C_i is the capital cost of the RES (USD), M_t is the annual maintenance cost of the RES (USD), L is the lifetime of the system (years) where it is assumed that all the components will have the same lifetime and r is the annual discount rate. The economic parameters used in the analysis are shown in Table 1.

Table 1. The photovoltaic (PV) and wind systems economic parameters and the grid tariff in addition to the yearly discount rate for Gwanda city, Zimbabwe.

Parameter	Value	Reference
Photovoltaic system capital cost (USD/kW)	1533	[58,59]
Wind system capital cost (USD/kW)	1516	[58,59]
Photovoltaic maintenance cost (USD/kW)	24.7	[60]
Wind maintenance cost (USD/kW)	39.53	[61]
System expected lifetime (Years)	25	[59,62]
Grid tariff (USD/MWh)	100	[38]
Annual discount rate (%)	7.2	[38]

3. Results and Discussion

Investment in renewable energy projects became attractive due to the profitability of such projects. Moreover, the deployment of RESs helps in the mitigation of greenhouse gases (GHGs) and at the same time ensures sufficient and secure energy sources. However, the significant drawback of RESs is the intermittency of the energy production from these systems due to the nature of the renewable energy resources which causes a mismatch between the demand and the supply and also affects the economics of the RESs.

The hybridization of RESs can overcome the intermittency of the resources and increase the harmony between the supply and the demand up to a certain limit where RESs can work in a synergistic way. Moreover, hybridization increases the economic benefits gained by the RESs which makes it more attractive to invest in. As the case study of this paper, Gwanda has a significant potential of wind and solar energy where the hybridization of these two resources proves that this option is the most suitable to achieve not only the maximum economic benefits but also the maximum technical benefits represented by the RES fraction, Figure 2 shows the increase in the technical feasibility of the renewable energy systems with the increase in the PV and wind capacities. However, it can be depicted from Figure 2 that after certain capacities' threshold is reached, the technical feasibility (represented by the RES fraction and the DSF) reaches saturation due to the mismatch between the demand and the supply which increase the electricity cost as shown in Figure 3. Hence, it vital to find the optimal PV and wind capacities that achieve the maximum technical and economic benefits, Table 2 shows the economic parameters as well as technical parameters of the maximum feasible capacities of PV, wind and PV-wind hybrid systems in Gwanda.

Table 2. The maximum feasible wind, PV and PV-wind hybrid systems capacities in Gwanda, Zimbabwe, in addition to their economic and technical parameters.

Configuration	Wind	PV	PV-Wind
PV Capacity (MW)	-	1.41	1
Wind Capacity (MW)	2	-	2
Capacity Factor (%)	34.57	18.72	29.29
RES Fraction (%)	54.99	28.72	65.07
DSF (%)	32.55	9.57	42.03
LCOE (USD/kWh)	0.09	0.10	0.10
NPV (million USD)	3.00	0.087	3.06
IRR (%)	17	7.65	14.04
PBP (years)	5	13.1	3.94

Note that in Table 2, the hybrid system does not have the highest RES fraction and DSF only, i.e., the highest technical feasibility but also it has the highest NPV and the lowest payback period (PBP) which means that the hybrid system in Gwanda achieves the maximum economic and technical benefits compared with separate systems where solar and wind resources complement each other in a synergistic way. Moreover, note that the PV system alone has the lowest RES fraction, DSF and the lowest economic benefits since it has the lowest capacity factor among the other options due to the nature of the solar resources where on average the PV meets totally the demand of Gwanda for two hours only in a day while the wind system on average meets the demand for five hours. On the other hand, the hybrid PV/wind system meets on average the demand nine hours a day. Figure 4 shows the average hourly demand of Gwanda as well as the average hourly energy generation from the maximum feasible PV, wind and PV-wind hybrid systems.

Note that in Figure 4, the hybridization between solar and wind systems achieves synergistic performance that increases the fraction of demand met by the RES and increases the autonomy of the RES system. Figure 5 shows the monthly RES fraction and demand supply fraction of the maximum feasible PV, wind and PV-wind systems in Gwanda.

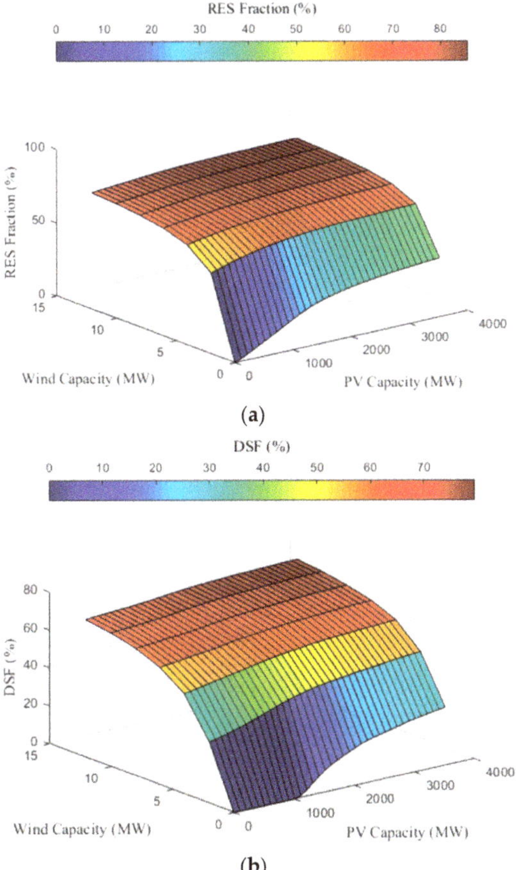

Figure 2. The technical parameters of different PV/wind hybrid system sizes in Gwanda: (**a**) renewable energy systems (RES) fraction and (**b**) demand supply fraction.

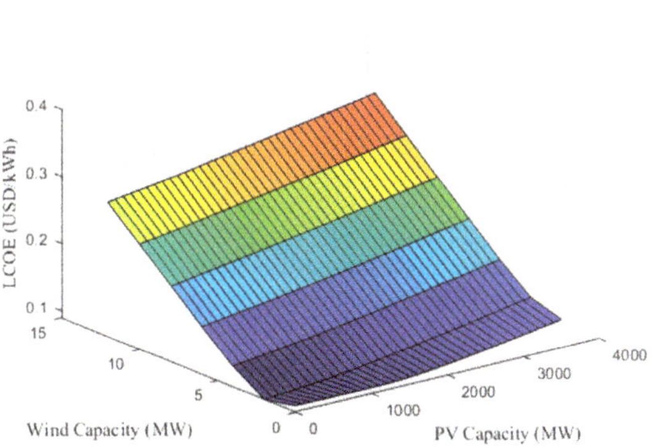

Figure 3. The levelized cost of electricity (LCOE) of different PV/wind hybrid system sizes in Gwanda.

Table 3. Techno-economic comparison of different systems at similar geographic locations in Africa.

Paper	Location	System Type	LCOE (USD/kWh)	NPV (USD)	RES Fraction (%)
[63]	South Africa	Standalone PV	0.16	-	-
[44]	South Africa	PV/CSP	0.16	-	-
[43]	Sub-Saharan Africa	Standalone PV	0.83	-	-
[64]	Nigeria	Standalone PV	0.40	590×10^3	-
[13]	Zimbabwe	Standalone PV	0.10	29.3×10^6	-
[38]	Zimbabwe	Hybrid Wind/PV	0.21	-	42
[15]	Zimbabwe	Hybrid PV/Wind-Battery	0.10	39.13×10^3	60.47
[65]	Zimbabwe	Wind System	0.13	7.8×10^6	-
This study	Zimbabwe	Hybrid PV/Wind	0.1	3.06×10^6	65.07

4. Conclusions

The hybridization between solar and wind systems partially solve the mismatch between the demand and the energy generation caused by the intermittency and the fluctuation of the resources and so increases the economic benefits of the RESs which makes it an attractive option to invest in. Solar and wind resources can complement each other and work in a synergistic way. Therefore, in order to prove that PV-wind hybrid system has better economics and performance compared with separate PV and wind systems, Gwanda city in Zimbabwe was the case study of this paper. The results indicate that the maximum feasible PV system in Gwanda-1.41 MW-has a RES fraction of 28.72% and NPV of 87×10^3 USD while the maximum feasible wind system-2 MW-has a RES fraction of 55% and NPV of 3×10^6 USD. On the other hand, the maximum feasible PV/wind system-2 MW wind and 1 MW PV-has RES fraction of 65.07% and NPV of 3.06×10^6 USD where it can be concluded that the PV-wind system did not only have the best technical performance but also the best economic benefit represented by the NPV and the PBP.

Author Contributions: Conceptualization, L.A.-G., R.S. and O.T.; methodology, L.A.-G. and O.T.; software L.A.-G.; validation, L.A.-G.; formal analysis, L.A.-G.; investigation, L.A.-G. and R.S.; resources, O.T. and M.F.; data curation, L.A.-G.; writing—original draft preparation, L.A.-G. and R.S.; writing—review and editing, L.A.-G., R.S., O.T. and M.F.; visualization, L.A.-G.; supervision, O.T. and M.F.; project administration, L.A.-G., O.T. and M.F. All authors have read and agreed to the published version of the manuscript.

Funding: This research received no external funding.

Acknowledgments: The authors would like to thank Derek Baker for providing the TMY data for Gwanda.

Conflicts of Interest: The authors declare no conflict of interest.

References

1. Al-Ghussain, L. Global warming: Review on driving forces and mitigation. *Environ. Prog. Sustain. Energy* **2018**, *38*, 13–21. [CrossRef]
2. Al-Ghussain, L.; Al-Oran, O.; Lezsovits, F. Statistical Estimation of Hourly Diffuse Radiation Intensity of Budapest City. *Environ. Prog. Sustain. Energy* **2020**. [CrossRef]
3. Hosseini, S.E. An outlook on the global development of renewable and sustainable energy at the time of COVID-19. *Energy Res. Soc. Sci.* **2020**, *68*, 101633. [CrossRef]
4. Wilson, J.E.; Grib, S.W.; Ahmad, A.D.; Renfro, M.W.; Adams, S.A.; Salaimeh, A.A. Study of Near-Cup Droplet Breakup of an Automotive Electrostatic Rotary Bell (ESRB) Atomizer Using High-Speed Shadowgraph Imaging. *Coatings* **2018**, *8*, 174. [CrossRef]
5. Ahmad, A.D.; Abubaker, A.M.; Salaimeh, A.A.; Akafuah, N.K. Schlieren Visualization of Shaping Air during Operation of an Electrostatic Rotary Bell Sprayer: Impact of Shaping Air on Droplet Atomization and Transport. *Coatings* **2018**, *8*, 279. [CrossRef]
6. Darwish Ahmad, A.; Singh, B.B.; Doerre, M.; Abubaker, A.M.; Arabghahestani, M.; Salaimeh, A.A.; Akafuah, N.K. Spatial Positioning and Operating Parameters of a Rotary Bell Sprayer: 3D Mapping of Droplet Size Distributions. *Fluids* **2019**, *4*, 165. [CrossRef]
7. Najjar, Y.S.H.; Abubaker, A.M. Exergy analysis of a novel inlet air cooling system with gas turbine engines using cascaded waste-heat recovery. *Int. J. Exergy* **2017**, *22*, 183–204. [CrossRef]

8. Abubaker, A.M.; Najjar, Y.S.H.; Ahmad, A.D. A Uniquely Finned Tube Heat Exchanger Design of a Condenser for Heavy-Duty Air Conditioning Systems. *Int. J. Air Cond. Refrig.* **2020**, *28*, 2050004. [CrossRef]
9. Najjar, Y.S.H.; Abubaker, A.M.; El-Khalil, A.F.S. Novel inlet air cooling with gas turbine engines using cascaded waste-heat recovery for green sustainable energy. *Energy* **2015**, *93*, 770–785. [CrossRef]
10. Najjar, Y.S.H.; Abubaker, A.M. Using novel compressed-air energy storage systems as a green strategy in sustainable power generation—A review. *Int. J. Energy Res.* **2016**, *40*, 1595–1610. [CrossRef]
11. Najjar, Y.S.H.; Abubaker, A.M. Indirect evaporative combined inlet air cooling with gas turbines for green power technology. *Int. J. Refrig.* **2015**, *59*, 235–250. [CrossRef]
12. Ahmad, A.D.; Abubaker, A.M.; Najjar, Y.S.H.; Manaserh, Y.M.A. Power boosting of a combined cycle power plant in Jordan: An integration of hybrid inlet cooling & solar systems. *Energy Convers. Manag.* **2020**, *214*, 112894.
13. Samu, R.; Fahrioglu, M. An analysis on the potential of solar photovoltaic power. *Energy Sources Part B Econ. Plan. Policy* **2017**, *12*, 883–889. [CrossRef]
14. Jingura, R.M.; Matengaifa, R. Rural energy resources and agriculture's potential as an Energy producer in Zimbabwe. *Energy Sources Part B Econ. Plan. Policy* **2009**, *4*, 68–76. [CrossRef]
15. Al-Ghussain, L.; Samu, R.; Fahrioglu, M. Techno-Economic Feasibility of PV/Wind-Battery Storage: Case Analysis in Zimbabwe. In Proceedings of the 16th International Conference on Clean Energy (ICCE-2018), Famagusta, Cyprus, 9–11 May 2018; pp. 9–11.
16. Al-Ghussain, L.; Samu, R.; Taylan, O.; Fahrioglu, M. Techno-Economic Analysis of Photovoltaic-Hydrogen Fuel Cell/Pumped Hydro Storage System for Micro Grid Applications: Case Study in Cyprus. In Proceedings of the 2018 International Conference on Photovoltaic Science and Technologies (PVCon), Ankara, Turkey, 4–6 July 2018; pp. 1–6.
17. Al-Ghussain, L.; Ahmed, H.; Haneef, F. Optimization of hybrid PV-wind system: Case study Al-Tafilah cement factory, Jordan. *Sustain. Energy Technol. Assess.* **2018**, *30*, 24–36. [CrossRef]
18. Al-Ghussain, L.; Taylan, O. Sizing methodology of a PV/wind hybrid system: Case study in cyprus. *Environ. Prog. Sustain. Energy* **2019**, *38*, e13052. [CrossRef]
19. Al-Ghussain, L.; Taylan, O.; Baker, D.K. An investigation of optimum PV and wind energy system capacities for alternate short and long-term energy storage sizing methodologies. *Int. J. Energy Res.* **2018**, 1–15. [CrossRef]
20. Al-Ghussain, L.; Samu, R.; Taylan, O.; Fahrioglu, M. Sizing Renewable Energy Systems with Energy Storage Systems in Microgrids for Maximum Cost-Efficient Utilization of Renewable Energy Resources. *Sustain. Cities Soc.* **2020**, *55*, 102059. [CrossRef]
21. Abujubbeh, M.; Marazanye, V.T.; Qadir, Z.; Fahrioglu, M.; Batunlu, C. Techno-Economic Feasibility Analysis of Grid-Tied PV-Wind Hybrid System to Meet a Typical Household Demand: Case Study-Amman, Jordan. In Proceedings of the 2019 1st Global Power, Energy and Communication Conference (GPECOM), Nevsehir, Turkey, 12–15 June 2019; pp. 418–423.
22. Rashid, M.; Abujubbeh, M.; Fahrioglu, M. Improving capacity factor of transmission lines by hybridizing CSP with wind. In Proceedings of the 2017 4th International Conference on Electrical and Electronic Engineering (ICEEE), Ankara, Turkey, 8–10 April 2017.
23. Abujubbeh, M.; Fahrioglu, M. Determining Maximum Allowable PV Penetration Level in Transmission Networks: Case Analysis-Northern Cyprus Power System. In Proceedings of the 2019 1st Global Power, Energy and Communication Conference (GPECOM), Nevsehir, Turkey, 12–15 June 2019; pp. 292–297.
24. Reiniger, K.; Schottland, T.; Zeidler, A. Optimization of Hybrid Stand-alone systems. In Proceedings of the European Wind Energy Association Conference and Exhibition, Rome, Italy, 7–9 October 1986.
25. Contaxis, G.C.; Kabouris, J. Short-term scheduling in a wind dieselautonomous energy system. *IEEE Trans. Power Syst.* **1991**, *6*, 1161–1167. [CrossRef]
26. Singh, G.; Baredar, P.; Singh, A.; Kurup, D. Optimal sizing and location of PV, wind and battery storage for electrification to an island: A case study of Kavaratti, Lakshadweep. *J. Energy Storage* **2017**, *12*, 78–86. [CrossRef]
27. Sadati, S.M.S.; Jahani, E.; Taylan, O.; Baker, D.K. Sizing of Photovoltaic-Wind-Battery Hybrid System for a Mediterranean Island Community Based on Estimated and Measured Meteorological Data. *J. Sol. Energy Eng.* **2018**, *140*, 011006. [CrossRef]

28. Azerefegn, T.M.; Bhandari, R.; Ramayya, A.V. Techno-economic analysis of grid-integrated PV/wind systems for electricity reliability enhancement in Ethiopian industrial park. *Sustain. Cities Soc.* **2020**, *53*, 101915. [CrossRef]

29. Bortolini, M.; Gamberi, M.; Graziani, A. Technical and economic design of photovoltaic and battery energy storage system. *Energy Convers. Manag.* **2014**, *86*, 81–92. [CrossRef]

30. Al-assad, R.; Ayadi, O. Techno-Economic Assessment of Grid Connected Photovoltaic Systems in Jordan. In Proceedings of the 2017 8th International Renewable Energy Congress (IREC), Amman, Jordan, 21–23 March 2017; pp. 1–4.

31. Pradhan, S.R.; Sahoo, S.P.; Das, R.; Priyanka, S. Design of Off-Grid Home with Solar-Wind-Biomass Energy. *Int. J. Eng. Res. Appl.* **2014**, *4*, 76–81.

32. Mehrpooya, M.; Mohammadi, M.; Ahmadi, E. Techno-economic-environmental study of hybrid power supply system: A case study in Iran. *Sustain. Energy Technol. Assess.* **2018**, *25*, 1–10. [CrossRef]

33. Jurasz, J.; Campana, P.E. The potential of photovoltaic systems to reduce energy costs for office buildings in time-dependent and peak-load-dependent tariffs. *Sustain. Cities Soc.* **2019**, *44*, 871–879. [CrossRef]

34. El-Tous, Y.; Al-Battat, S.; Abdel, H.S. Hybrid Wind-PV grid connected Power Station Case Study: Al-Tafila, Jordan. *Int. J. Energy Environ.* **2012**, *3*, 605–616.

35. Al-Masri, H.; Amoura, F. Feasibility Study of A Hybrid Wind/PV System Connected to the Jordanian Grid. *Int. J. Appl. Power Eng.* **2013**, *2*. [CrossRef]

36. Benlouba, S.; Bourouis, M. Feasibility Study of a Wind-Photovoltaic Power generation System for a remote area in the extreme south of Algeria. *Appl. Energy* **2016**, *99*, 713–719. [CrossRef]

37. Ashok, S. Optimised model for community-based hybrid energy system. *Renew. Energy* **2007**, *32*, 1155–1164. [CrossRef]

38. Samu, R.; Fahrioglu, M.; Taylan, O. Feasibility Study of a Grid Connected Hybrid PV-Wind Power Plant in Gwanda, Zimbabwe. In Proceedings of the IEEE Honet Symposium, Nicosia, Cyprus, 13 October 2016; pp. 122–126.

39. Asumadu-Sarkodie, P.A.; Owusu, S. The potential and economic viability of solar photovoltaic power in Ghana. *Energy Sources Part A* **2016**, *38*, 709–716. [CrossRef]

40. Adaramola, M. Viability of grid-connected solar PV energy system in Jos, Nigeria. *Int. J. Electr. Power Energy Syst.* **2014**, *61*, 64–69. [CrossRef]

41. Kebede, K.Y. Viability study of grid-connected solar PV system in Ethiopia. *Sustain. Energy Technol. Assess.* **2015**, *10*, 63–70. [CrossRef]

42. L-Shimy, M.E. Viability analysis of PV power plants in Egypt. *Renew. Energy* **2009**, *34*, 2187–2196. [CrossRef]

43. Baurzhan, S.; Jenkins, G. Off-grid solar PV: Is it an affordable or appropriate solution for rural electrification in Sub-Saharan African countries? *Renew. Sustain. Energy Rev.* **2016**, *60*, 1405–1418. [CrossRef]

44. Pan, C.A.; Dinter, F. Combination of PV and central receiver CSP plants for base load power generation in South Africa. *Sol. Energy* **2017**, *146*, 379–388. [CrossRef]

45. Chahuruva, R.; Dei, T. Study on Isolated Solar Home Systems for Application in Zimbabwe. *Energy Procedia* **2017**, *138*, 931–936. [CrossRef]

46. Ziuku, S.; Seyitini, L.; Mapurisa, B.; Chikodzi, D.; van Kuijk, K. Potential of Concentrated Solar Power (CSP) in Zimbabwe. *Energy Sustain. Dev.* **2014**, *23*, 220–227. [CrossRef]

47. Mentis, D.; Hermann, S.; Howells, M.; Welsch, M.; Siyal, S. Assessing the technical wind energy potential in Africa a GIS-based approach. *Renew. Energy* **2015**, *83*, 110–125. [CrossRef]

48. Fant, C.; Gunturu, B.; Schlosser, A. Characterizing wind power resource reliability in southern Africa. *Appl. Energy* **2016**, *161*, 565–573. [CrossRef]

49. Bogno, B.; Sali, M.; Aillerie, M. Technical and Economic Analysis of a Wind Power Generation System for Rural Electrification in Subequatorial Area of Africa. *Energy Procedia* **2014**, *50*, 773–781. [CrossRef]

50. Kazet, M.; Mouangue, R.; Kuitche, A.; Ndjaka, J.M. Wind Energy Resource Assessment in Ngaoundere Locality. *Energy Procedia* **2016**, *93*, 74–81. [CrossRef]

51. Ministry of Energy and Power Development. *National Renewable Energy Policy*; Ministry of Energy and Power Development: Harare, Zimbabwe, 2019.

52. Dubey, S.; Sarvaiya, J.; Seshadri, B. Temperature dependent photovoltaic (PV) efficiency and its effect on PV production in the world—A review. *Energy Procedia* **2013**, *33*, 311–321. [CrossRef]

53. Canadian Solar. *Photovoltaic Panels Datasheet*; Canadian Solar: Guelph, ON, Canada, 2017.

54. Duffie, J.; Beckman, W. *Solar Engineering of Thermal Processes*, 3rd ed.; Wiley: Hoboken, NJ, USA, 2006.
55. Manwell, J.F.; McGowan, J.G.; Rogers, A.L. *Wind Energy Explained: Theory, Design and Application*, 2nd ed.; Wiley: Chichester, UK, 2009.
56. Gamesa. *Greater Energy Produced From Low and Medium Wind Sites*; Gamesa: Zamudio, Spain, 2014.
57. Tembo, B.; Merven, B. Policy options for the sustainable development of Zambia's electricity sector. *J. Energy S. Afr.* **2013**, *24*, 16–27. [CrossRef]
58. Fichter, T.; Trieb, F.; Moser, M.; Kern, J. Optimized integration of renewable energies into existing power plant portfolios. *Energy Procedia* **2013**, *49*, 1858–1868. [CrossRef]
59. Al-Ghussain, L.; Taylan, O.; Fahrioglu, M. Sizing of a Photovoltaic-Wind Oil Shale Hybrid System: Case Analysis in Jordan. *J. Sol. Energy Eng. Incl. Wind Energy Build. Energy Conserv.* **2018**, *140*, 1–12. [CrossRef]
60. Sangster, A.J. Solar Photovoltaics. *Green Energy Technol.* **2014**, *194*, 145–172.
61. Breeze, P. Wind Power. *Power Gener. Technol.* **2014**, *1*, 223–242.
62. Yang, H.; Wei, Z.; Chengzhi, L. Optimal design and techno-economic analysis of a hybrid solar-wind power generation system. *Appl. Energy* **2009**, *86*, 163–169. [CrossRef]
63. Numbi, B.P.; Malinga, S.J. Optimal energy cost and economic analysis of a residential grid-interactive solar PV system-case of eThekwini municipality in South Africa. *Appl. Energy* **2017**, *186*, 28–45. [CrossRef]
64. Okoye, C.O.; Oranekwu-Okoye, B.C. Economic feasibility of solar PV system for rural electrification in Sub-Sahara Africa. *Renew. Sustain. Energy Rev.* **2018**, *82*, 2537–2547. [CrossRef]
65. Samu, R.; Fahrioglu, M.; Ozansoy, C. The potential and economic viability of wind farms in Zimbabwe. *Int. J. Green Energy* **2019**, 1–8. [CrossRef]

Article

Full-Scale Implementation of RES and Storage in an Island Energy System

Konstantinos Fiorentzis, Yiannis Katsigiannis and Emmanuel Karapidakis *

Department of Electrical and Computer Engineering, Hellenic Mediterranean University, GR-71004 Heraklion, Greece; kfiorentzis@hmu.gr (K.F.); katsigiannis@hmu.gr (Y.K.)
* Correspondence: karapidakis@hmu.gr; Tel.: +30-2810-379-889

Received: 10 September 2020; Accepted: 29 October 2020; Published: 30 October 2020

Abstract: The field of energy, specifically renewable energy sources (RES), is considered vital for a sustainable society, a fact that is clearly defined by the European Green Deal. It will convert the old, conventional economy into a new, sustainable economy that is environmentally sound, economically viable, and socially responsible. Therefore, there is a need for quick actions by everyone who wants to move toward energy-efficient development and new environmentally friendly behavior. This can be achieved by setting specific guidelines of how to proceed, where to start, and what knowledge is needed to implement such plans and initiatives. This paper seeks to contribute to this very important issue by appraising the ability of full-scale implementation of RES combined with energy storage in an island power system. The Greek island power system of Astypalaia is used as a case study where a battery energy storage system (BESS), along with wind turbines (WTs), is examined to be installed as part of a hybrid power plant (HPP). The simulation's results showed that the utilization of HPP can significantly increase RES penetration in parallel with remarkable fuel cost savings. Finally, the fast response of BESS can enhance the stability of the system in the case of disturbances.

Keywords: energy transition; renewable energy sources; island power systems; hybrid power plants; wind turbines; battery energy storage systems

---

1. Introduction

The integration of renewable energy sources (RES) is interesting to designers of isolated island power systems, presenting significant opportunities especially for fuel cost savings. Islands are usually very dependent on fossil fuel imports, which are typically expensive due to transport costs. Hence, the utilization of RES via the exploitation of the island's RES potential can assist the load demand–supply, reducing the cost derived from fuel consumption, as well as pollutant emissions [1,2]. Two strategies have been developed by the European Union (EU) in view of climate change and sustainable development on islands [3,4] with the aim to maximize the share of RES. In this respect, such island power systems seek to improve their independence from conventional units, promoting an environmentally friendly profile.

However, RES integration traditionally brings operating issues in power systems. Due to their strong weather dependence, such as in the cases of solar irradiation and wind speed, RES generation can have large fluctuations. These unpredictable variations affect the stability and power quality of power systems. These issues are primarily related to voltage and frequency deviations [5]. In comparison with interconnected power systems, such issues are increased in isolated island communities, as fluctuations in production or demand changes can lead to larger frequency deviations [6,7]. Therefore, to avoid problems that can affect the safety and stability of the power supply system, intermittent RES power has to be limited to a higher specific percentage of the system's load, compared with interconnected power systems. These actions arise from the need for ensuring adequate frequency and voltage regulation, which are traditionally provided by conventional synchronous units. Subsequently, in scenarios of high RES penetration, the operator of the system may give the command of RES curtailment with respect to the operational requirements of conventional units and the spinning reserve criteria [8].

A solution in RES power curtailment is the utilization of energy storage systems (ESSs). ESSs are mainly used to maximize the RES penetration [9–13] by storing RES power/energy in periods that cannot be absorbed by the power systems, providing it in periods when it is required. Furthermore, ESSs can be integrated as a technique to maintain stability in cases of power system stress. Due to their fast response, ESSs can be used from the operator of a system to regulate frequency [14,15] and voltage [16,17], maintaining them at the desired limits, either via power injection (frequency/voltage increase) or power absorption (frequency/voltage decrease). Therefore, the utilization of RES in combination with energy storage systems can lead to near-zero or 100% RES island power systems. In such systems, the dependence on conventional units is almost eliminated, resulting in them practically being used as backup systems for ancillary service provision [18].

Creek territory consists of 32 isolated island power systems with the vast majority of them in the Aegean Archipelagos [19]. During the last decade, a procedure began to develop a sea transmission network that contains all insular systems. This network is planned to be fully developed by the end of the next decade. Most of these isolated power systems present significant wind and solar potential. Hence, considerable amounts of RES penetration levels have been reached. By 2017, 97 wind farms (323 MW) were in operation on all Greek noninterconnected insular systems in addition to 758 PV parks (136 MW), 242 rooftop PV systems (24 MW), and one small hydroelectric station (0.3 MW) [20]. Additionally, extensive research has been carried out on large-scale energy storage development in those islands. The operation of wind-pump storage units in the Cretan power system was examined in [21,22], while, in [23,24], the impact of hybrid power systems was evaluated for the Samos island power system; in [25], a hybrid power plant was utilized for Sifnos island to reach 100% energy autonomy. Currently, Ikaria island is the only Greek isolated island with a large-scale hybrid power system in operation [26].

This paper investigates the technical and economical optimal generation scheme in a specific real grid that belongs to the small Greek island power system of Astypalaia, utilizing a hybrid power plant (HPP). Taking into account similar research studies [24,27–29], isolated island power systems, such as that of Astypalaia which presents significant wind potential, could be representative case studies for further RES implementation, not only in European isolated islands but also worldwide. The utilization of such a large-scale HPP represents an energy planning strategy for further RES implementation on islands, which contributes to the full-scale deployment of green power technologies. More specifically, different scenarios of battery energy storage systems (BESSs) and wind turbines (WTs) are combined to significantly reduce the operation of the existing conventional units of the island. Therefore, an extensive analysis is made presenting annual data for the electricity production of the existing thermal units and the BESS; the purchased annual energy needed for BESS charging, as well as the annual discarded WT energy and WT energy injection to the grid, is also calculated. Furthermore, the annual fuel costs savings for each examined scenario are presented while an extensive economic analysis is also made using financial indicators. Finally, a stability analysis for the examined power system is conducted, after the installation of the HPP.

This paper is organized as follows: Section 2 presents the most significant features and restrictions of the Greek legislation framework regarding the operation of HPPs in insular power systems. Section 3 contains a general description of the island power system of Astypalaia, Section 4 contains the methodology adopted, while Section 5 contains the main simulation results and the economic analysis for HPP utilization. Section 6 contains the analysis for the dynamic performance of the power system after the installation of the HPP, while Section 7 concludes the paper.

2. Greek Legislation for Hybrid Power Plants in Insular Power Systems

In general, hybrid power plants (HPP) consist of a storage system, a controllable generation unit, and at least one form of RES power generation unit. A wide range of technologies can be used for an HPP's electricity storage and production. Pumped hydro storage units and battery stations represent the most common and mature technologies for large-scale energy storage, in terms of HPP. In Greece, according to [30], a power plant is defined as an HPP if it meets the following requirements/criteria:

- The plant comprises at least one RES unit and a storage system;
- The total electricity absorbed from the grid, on a yearly basis, does not exceed 30% of the total stored electrical energy;
- The maximum installed capacity of the HPP's RES units does not exceed the respective installed capacity of the storage units, increased by 20%.

The installation of a wind-battery HPP on the Greek noninterconnected insular power system (NIIS) of Astypalaia is examined in this paper. According to [31], the operator of the NIIS is obliged to absorb the electricity produced by the RES with priority, including the HPP's units, over conventional units without prejudice to the secure operation of the NIIS. Within this frame, this priority is not valid when the HPP's production violates the restriction for the technical minimum of the must-run conventional units. Additionally, there is not a priority in HPP's commitment over conventional units when their production is deemed necessary to meet ancillary service requirements which are not possible to be met by RES and HPPs.

In terms of HPP utilization from the operator of the NIIS, the declaration of its guaranteed energy, which is equal to the product of maximum battery capacity P_{bat}, and the number of hours of guaranteed power are required. Regarding the operational principle of a wind-battery HPP, two basic case scenarios can be considered:

- Case 1: if the total power output of the wind farms is less than the installed capacity of HPP P_{bat}, the total generated wind power can be stored in the HPP with respect to the battery minimum and maximum state of charge (SoC).
- Case 2: if the total power output of the wind farms is greater than P_{bat} and less than $1.2 \times P_{bat}$, the amount of wind power that cannot be stored can be provided directly to the grid, in the case that there is the capability of additional power injection to the grid from RES; otherwise, it is discarded.

Other restrictions related to HPP operation include that (a) the provided energy from the HPP the first 12 h of the day cannot exceed the provided energy of the last 12 h of the day, (b) the daily produced energy has to be at least $2h \times P_{bat}$ (otherwise must be equal to zero), (c) on certain days (especially with high loads), the HPP has to provide its guaranteed energy, and (d) the price of the electricity taken from the battery station is higher compared to the price of electricity sold by wind farms to the grid.

3. Description of Astypalaia Island Autonomous Power System

Astypalaia is a small island that belongs to the Dodecanese complex in the southeastern Aegean Sea, Greece. Its permanent population is 1334 inhabitants (2011 data); however, during summer months there exists a significant increase, which is depicted in load consumption. The annual peak load reached 2.22 MW (2014 data) on 15 August at 9:00 p.m. Currently, the autonomous power system of Astypalaia island is fed by a diesel power station that consists of three identical diesel generators (Mitsubishi S16R-PTA) with 1 MW peak power each (3 MW in total). Moreover, 320 kWp of photovoltaic (PV) power has been installed on the island, with annual electricity production of 531.38 MWh in 2019 (18.96% annual capacity factor) [32]. Figure 1a shows the net load duration curve, which is the full load minus PV production, and Figure 1b depicts the annual load demand, which shows the significant load increase (more than double in some periods) during summer months.

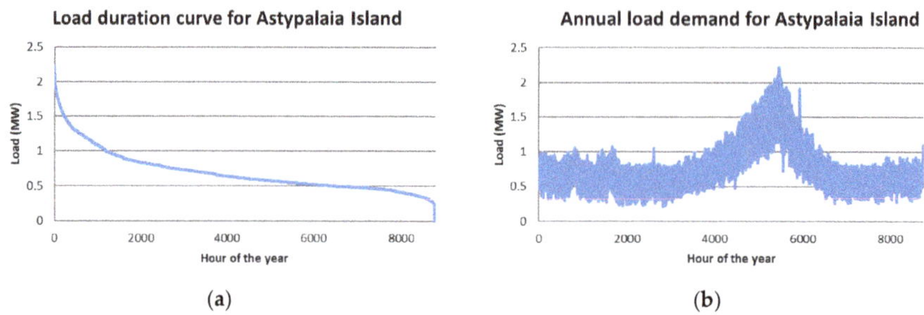

Figure 1. Net load on Astypalaia Island for the year 2014: (**a**) load duration curve; (**b**) annual load demand.

The island of Astypalaia presents significant wind potential, as with the majority of locations in the Dodecanese complex. Although no wind turbines have been installed, there is a large number of approved wind farms (by the Greek Regulatory Authority for Energy) for electricity generation on the island (on the scale of several tenths of megawatts) [33]. For a typical mountainous location on Astypalaia Island, although the annual mean wind speed is around 6.5–7.0 m/s (for a 10 m anemometer height), there are no high wind speed values that surpass 20 m/s. This characteristic makes this location ideal for a WT that is suitable for medium to high wind speeds. The proposed WT for installation is the Siemens Gamesa onshore model SG 2.6-114, with 2.625 MW rated power, a hub height of 88 m (other hub height alternatives are also available), and wind class IEC IA/IIA/S [34], which corresponds to medium and high wind speeds [35]. Figure 2a,b show the mean monthly wind speed and wind speed duration curve, respectively, taken from a mountainous location in the northwest of Astypalaia, using typical model year (TMY) data provided in [36]. Figure 3 shows the power curve of the SG 2.6-114 WT model.

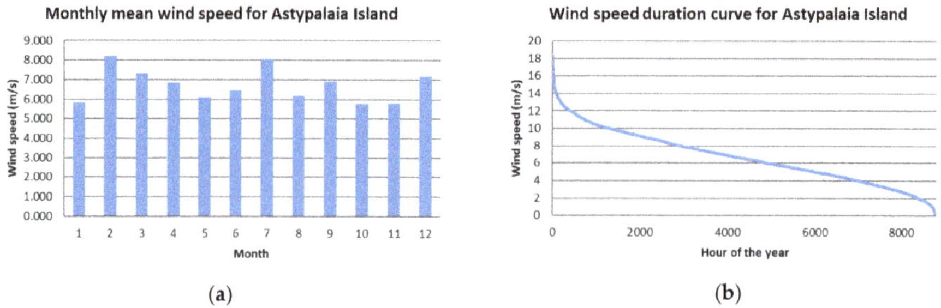

Figure 2. Wind data for a mountainous location in the northwest of Astypalaia (10 m anemometer height with annual mean wind speed of 6.73 m/s): (**a**) monthly mean wind speed; (**b**) wind speed duration curve.

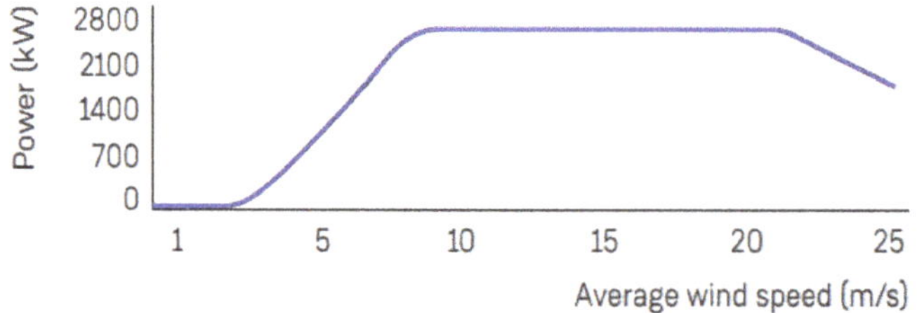

Figure 3. Power curve of SG 2.6-114 wint turbine (WT) model [34].

4. Methodology

In the following analysis, time-series data with an hourly time step (i.e., 8760 annual values) were used, which include net load, total WT production, and total PV production. WT production was calculated by combining the annual wind data (Figure 2) and WT power curve (Figure 3). Moreover, the effect of WT hub height was also taken into account using the power law with a power law exponent value a = 1/7 [37]. The annual capacity factor for each WT was equal to 41.03% (around 3600 equivalent hours per year operating at rated power). PV production data were used only in unit commitment calculations (see Equation (1)) and were estimated by PV-GIS [36]. The difference between the annual estimated and observed PV electricity production was only 0.8%.

Initially, a simulation of the system's operation in its current state (without HPP installation) was implemented. For the unit commitment problem, the priority list method was used for the three identical diesel generators. A strict rule for spinning reserve was considered as follows [22]:

$$\sum_i u_i \cdot P_{imax} \geq 1.1 \cdot P_{Load} + 0.2 \cdot P_{WT} + 0.1 \cdot P_{PV}, \tag{1}$$

where u_i is the i unit status (1 for on and 0 for off), P_{imax} is the maximum power of unit i, P_{Load} is the net load demand, P_{WT} is the WT production, and P_{PV} is the PV production. As a first step, a simulation run was executed for the annual operation of Astypalaia island power system, without considering the installation of the HPP. For each hour, the load to be supplied by diesel generators was equal to the net load minus WT production ($P_{Load} - P_{WT}$). Considering the technical data of diesel generators and the spinning reserve constraint described in Equation (1), the hourly production for each unit could be calculated.

Next, the operation of Astypalaia Island operation was evaluated considering the installation of the HPP units. Each HPP unit consisted of a combination of Siemens Gamesa onshore WT model SG 2.6-114 and a Narada lead-carbon BESS of 1 MW alternating current (AC) output power [34]. In each case, multiples of this combination could be used. The technical characteristics of Narada BESS were as follows:

- Output energy (AC): typically 8 MWh, whereas other capacities could be used (usually in the range of 4–12 MWh);
- Single-way efficiency equal to 0.96, which means that the roundtrip efficiency (n_{total}) was equal to 0.9216;
- Typical battery lifetime: 5000 (daily) cycles (more than 13.5 years).

The basic criterion for choosing lead–carbon batteries for the proposed BESS of the HPP was economic. According to brief calculations, the utilization of lithium-ion batteries would have given a nonviable solution for the proposed investment, as they are significantly more expensive than lead–carbon batteries. Therefore, it was preferable to use lead–carbon batteries which also promote a well-proven technology as they are the next generation of the older lead–acid technology. Furthermore, this paper proposes the integration of an HPP with the aim of providing an indicative energy planning strategy for the energy transition of island power systems, making them more energy-efficient, economically efficient, and environmentally friendly. Taking into account the restriction of an environmentally friendly power system, the advantage of lead–carbon batteries is that they can be recycled in contrast with lithium batteries, where the issue of their recycling remains an unresolved problem.

Regarding the operation of HPP, the WT electricity production for each hour of the year was initially calculated. The maximum WT penetration (concerning the net load) was considered to be 60%, such that if the WT production surpassed this limit for a specific hour, the surplus energy was curtailed. Then, the charging procedure of the BESS from WTs was estimated, according to the rules described in Cases 1 and 2 of Section 2. In cases when, during charging, the BESS SoC reached 100%, the WT energy that was used for BESS charging was reduced properly and it was absorbed from grid if possible.

The load demand of each day was assumed to be known (zero-load forecasting error); thus, during each day that the BESS was operating, its discharging schedule could be estimated by considering that the BESS discharged during peak load hours (usually evening and early night hours). As a result, peak load was reduced (peak-shaving). On days when the power system operator needed the guaranteed energy from the BESS, regardless of WT production, if the WT production was not sufficient to fully charge the BESS, power from the grid also had to be absorbed. This usually happened during night hours, when the load presented its minimum daily values (valley-filling). In this paper, this was considered to take place on all days of the year where the daily energy demand was greater than 80% of the maximum daily energy demand of the year. For the Astypalaia Island power system, this corresponded to 25 days per year (mainly in the summer period).

Taking into account the load reduction during WT production and peak-shaving, as well as the load increase during valley-filling, the new load curve could be estimated; thus, the operating points of diesel generators considering the HPP installation could be calculated using Equation (1). The flowchart of the abovementioned methodology is presented in Figure 4.

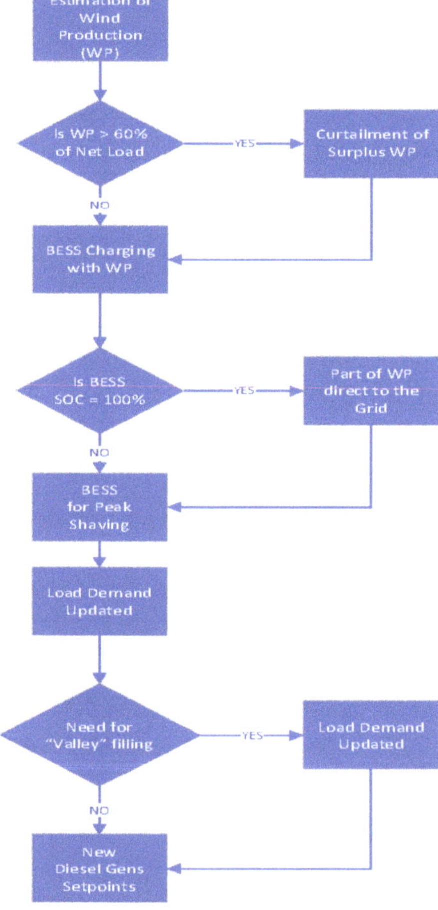

Figure 4. Flowchart of Astypalaia non-interconnected insular power system (NIIS) operation considering hybrid power plant (HPP) installation.

5. Economic Analysis, Results

The economic analysis was based on the directives of the Greek legislative framework for noninterconnected islands. The energy delivered from BESS to the grid had a cost of EUR 147/MWh. Energy absorbed from the grid had a cost of EUR 135.47/MWh, which was equal to EUR 147 × n_{total}/MWh. Electricity from the installed WTs that could not be absorbed by the BESS but was absorbed by the grid (if maximum wind penetration was not surpassed) had a cost of EUR 98/MWh. The considered initial costs were EUR 250,000/MWh for the BESS (without WTs) and EUR 1,200,000/MW for the WTs. Seventy percent of these initial costs were covered from a bank loan at a 7.5% interest rate with a 15 year duration. Total annual operational and maintenance (O&M) costs were assumed to be 1% of the initial cost. The annual discount rate i was 6%, whereas the total project lifetime was considered to be 25 years, which is equal to the WT lifetime. Batteries were replaced during the 14th year of system operation (5000 daily cycles). For WTs and BESSs, during the end of their lifetime, a salvage value equal to 20% of their initial cost was assumed.

For all developed scenarios, MATLAB software was used. The following financial indicators were calculated:

1. Net present value (NPV): positive NPV values (in Euros) are an indicator of a potentially feasible project; however, NPV also has to be compared with the size of the project (initial investment).
2. Internal rate of return (IRR): a project is acceptable if the IRR is greater than the discount rate ($i = 6\%$).
3. Benefit-to-cost (B/C) ratio: this is the ratio of the net benefits to costs of the project. B/C ratios greater than 1 are indicative of profitable projects.
4. Simple payback (in years): contrary to the previous indices, simple payback does not consider the time value of money. It is useful, however, as a secondary indicator to indicate the level of risk of an investment.

NPV was used as the primary financial indicator for the evaluation of HPP installation. All scenarios with two WTs led to large negative NPV values (see, for example, the last column of Table 1) and significantly higher amounts of discarded wind energy (approximately three times more for the same BESS capacity). Figure 5 shows the total NPV for HPP projects with one WT and different BESS capacity. The results show that the most beneficial case was the installation of the 8 MWh BESS. Table 1 shows the financial indicators for a number of considered scenarios, including a typical one that contains two WTs. All three indicators that took into account the time value of money showed that the optimal scenario was that highlighted in Figure 5.

Table 2 gives a more detailed analysis for the four scenarios with one WT that presented the highest (positive) NPVs. The parameters that were calculated include the total net load, the annual electricity production of diesel generators and BESS, the annual energy purchased from the grid (in the case of valley-filling at night hours), the annual wind energy fed directly to the grid (without being stored to BESS), and the annual discarded wind energy (i.e., the wind energy not absorbed by the BESS or the grid). The last two parameters of Table 2 provide information about the annual energy penetration of HPP in Astypalaia Island. The results show that the increase in BESS capacity decreased conventional generation and WT energy fed directly to the grid or discarded and increased the penetration of HPP. Although the increase in BESS capacity improved the operational characteristics of the studied insular system, the high initial cost of batteries did not lead to better financial performance of the HPP for very large BESS capacities.

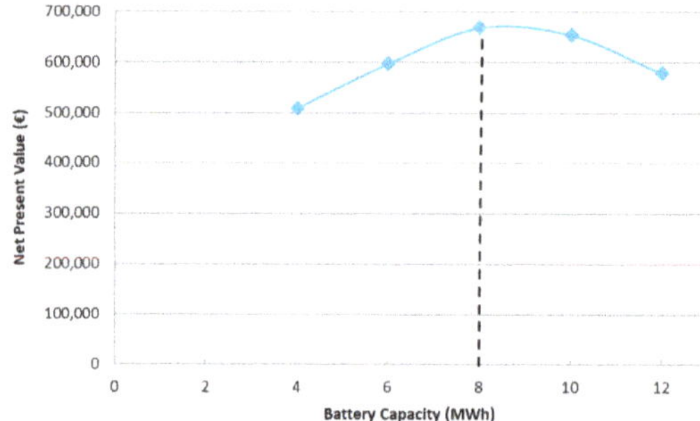

Figure 5. Total net present value (NPV) of HPPs containing one WT and different battery energy storage system (BESS) capacity.

Table 1. Financial indices for a number of considered scenarios. IRR, internal rate of return; B/C ratio, benefit-to-cost ratio.

Index	1 WT, 1MW, 6 MWh BESS	1 WT, 1MW, 8 MWh BESS	1 WT, 1MW, 10 MWh BESS	2 WTs, 2MW, 8 MWh BESS
NPV (EUR)	597,893	669,507	654,782	−1,829,710
IRR	8.49%	8.57%	8.32%	2.12%
B/C ratio	1.43	1.43	1.39	0.27
Simple payback (years)	17.62	17.80	18.17	24.07

Table 2. Annual electricity production and consumption for Astypalaia Island.

Parameter	1 WT, 1MW, 6 MWh BESS	1 WT, 1MW, 8 MWh BESS	1 WT, 1MW, 10 MWh BESS	1 WT, 1MW, 12 MWh BESS
Total load (MWh)	6188.43	6188.43	6188.43	6188.43
Diesel generators (MWh)	2089.25	1741.17	1427.55	1130.53
BESS (MWh)	2082.31	2690.96	3230.63	3706.77
Purchased from grid (valley-filling, in MWh)	9.75	9.76	9.75	9.91
WT energy directly provided to grid (MWh)	2026.61	1766.06	1540.00	1361.04
Discarded wind energy (MWh)	5157.86	4758.01	4398.47	4060.95
BESS penetration	33.65%	43.48%	52.20%	59.90%
BESS + WT penetration	66.40%	72.02%	77.09%	81.89%

Table 3 compares the annual fuel costs of diesel generators for all examined profitable scenarios, as well as for the current situation (operation without WTs or HPPs). Even the installation of an HPP with small BESS capacity significantly decreased (50% or more) annual fuel costs. Figure 6 shows the duration curves of BESS SoC for the four most dominant scenarios that were also included in Table 2. In the case of a small BESS capacity, there were several hours during the year in which the BESS was fully charged: more than 2800 h for the 6 MWh capacity (see Figure 6a) and more than 4000 h for the 4 MWh capacity. This usually happened on high-wind night and morning hours, during which the BESS was being charged by WTs. The small BESS capacity led to an inability to fully exploit the available wind potential, which also affected the financial indicators, significantly reducing revenues that could be received from the energy sold to the grid.

Table 3. Comparison of annual fuel costs for considered scenarios in Astypalaia Island.

Title 1	Annual Fuel Costs (EUR)
Current state	1,092,703
1 WT, 1MW, 4 MWh BESS	521,933
1 WT, 1MW, 6 MWh BESS	476,754
1 WT, 1MW, 8 MWh BESS	430,233
1 WT, 1MW, 10 MWh BESS	387,582
1 WT, 1MW, 12 MWh BESS	347,990

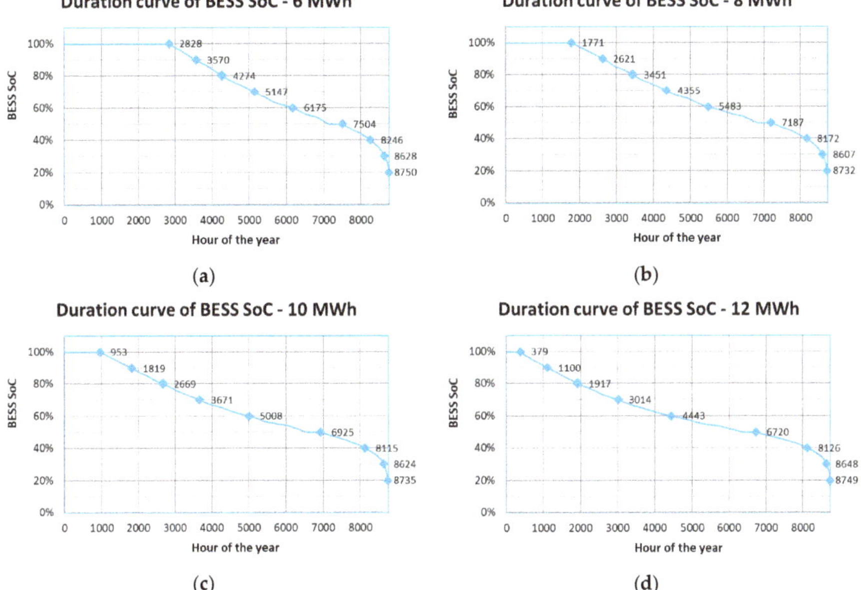

Figure 6. Duration curve for state of charge (SoC) for systems with 1 WT and different BESS capacity: (**a**) 6 MWh; (**b**) 8 MWh; (**c**) 10 MWh; (**d**) 12 MWh.

6. Dynamic Performance of the System with HPP

In this section, a preliminary stability analysis of the proposed production scheme was conducted. The profile of the HPP that was investigated through simulations (using Powerworld Simulator version 21 software) consisted of one WT of 2.625 MW and a BESS of 1 MW power and 8 MWh energy capacity. Previous power flows were used and basic disturbances were assessed. In more than 7000 h (cases A) of operation under the preselected specific disturbances, the island grid could be considered substantially safe, due to the fast response of the 1 MW BESS, while most of the remaining cases (cases B) retained their robustness under the precondition of sufficient wind power generation (equipped with Fault Ride Through (FRT)). Concluding, a few cases (cases C) with high load demand more than 1.5 MW and inadequate wind power production less than 0.3 MW were assessed as unsafe. This could be counterbalanced by the contribution of spinning reserves of the existing internal combustion units of the island. Therefore, the vital role of accurate real-time load and wind forecasting was once again confirmed.

In Figure 7, the variations in frequency profiles for several operation cases are depicted as described previously. The system was considered to operate with adequate energy storage power capacity (cases A) and with sufficient wind turbine generation (cases B), and it seemed to be quite stable. In cases C with insufficient wind production and storage power, the generation system could be supported by conventional units such as the current diesel generators (fast spinning reserve), which could clearly contribute to grid robustness.

Concluding, as a next step of this research, extensive simulation of the proposed HPP should be carried out to identify specific real-time operation guidelines and regulations, aiming at the most stable profile, which would cover all the frequency, voltage, and power quality requirements of weak grids as island power systems.

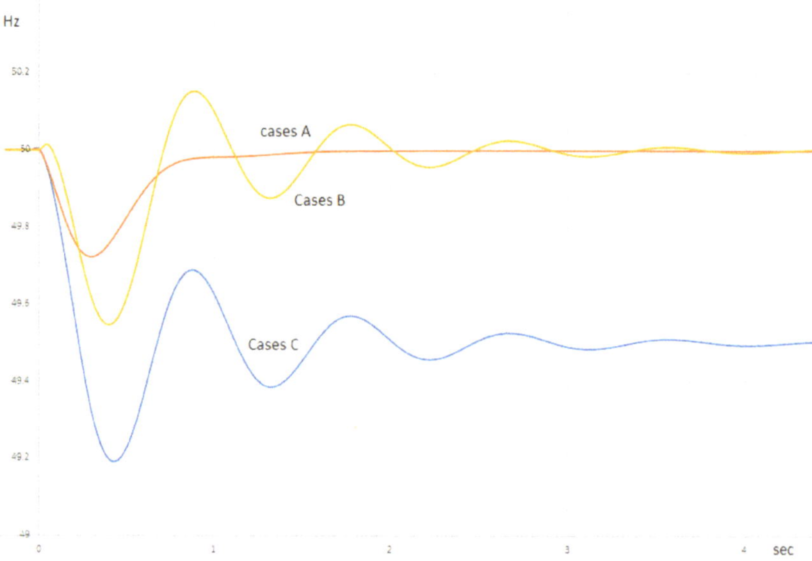

Figure 7. Frequency variations.

7. Conclusions

In this paper, the operation of the Greek insular power system of Astypalaia was examined, after the installation of an HPP consisting of WTs and BESSs. Due to the annual mean wind speed of 6.5–7.0 m/s observed on the island, a WT suited for medium to high wind speeds was selected to be installed. This WT was a Siemens Gamesa onshore model SG 2.6-114, with 2.625 MW rated power and 88 m hub height. Additionally, a Narada lead-carbon BESS of 1 MW AC output power was selected. For the needs of the simulation, combinations of those WTs and BESS were evaluated to find the optimal technical and especially economical generation scheme.

The criterion used to find this optimal scheme was the results of economic analysis using financial indicators. The relevant calculations were made considering the following initial costs: EUR 250,000/MWh for BESS (5000 daily cycles) and EUR 1,200,000/MW for WTs. Furthermore, the energy injected from the BESS to the grid had a cost of EUR 147/MWh, the cost of energy absorbed from the grid was assumed as EUR 135.47/MWh, and the total project lifetime was considered to be 25 years. Other considerations for the investment were that 70% of these initial costs were covered from a bank loan of 7.5% interest rate with a 15 year duration. The total annual O&M costs were assumed to be 1% of the initial cost, and the annual discount rate i was 6%. Taking into account those investment considerations, an analysis was made calculating the indicators of NPV, IRR, B/C ratio, and simple payback for different scenarios of BESS capacity with one SG 2.6-114 WT.

The results for all four indicators showed that the most beneficial combination was the installation of the 8 MWh BESS with one SG 2.6-114 WT. Consequently, a simulation was made to examine the annual energy penetration of the HPP in the Astypalaia island power system. Keeping as a basic scenario the utilization of one SG 2.6-114 WT, different scenarios of BESS capacity were examined for this purpose. The results showed that the increase in BESS capacity led to a decrease in the thermal unit's annual electricity production. A decrease was also observed in discarded WT energy or WT injection to the grid, while the penetration of HPP was also increased. However, due to the high initial cost of batteries, the increase in BESS capacity could not lead to a better financial performance of the HPP. In the case of a small BESS capacity (4 MWh and 6 MWh), it was almost impossible for wind potential to be fully exploited, thus reducing revenues that could be received from the energy sold to the grid. Thus, the capacity of 8 MWh BESS was confirmed as the optimal selection.

Finally, the stability analysis showed that the Astypalaia island power system could become much more robust after the installation of the examined HPP. More specifically, in more than 7000 h of operation in annual level, the island grid could be considered substantially safe, as the scheme of the 1 MW BESS would immediately respond to a possible disturbance. An unsafe situation was observed only in a few cases (less than 300 h annually) of load demand higher than 1.5 MW. In such cases, the spinning reserves of fast response units such as the existing internal combustion units could act to keep the robustness of the system.

As an overall conclusion, the island of Astypalaia was used as a case study with the aim of proposing an indicative energy planning strategy for the energy transition of isolated and weak interconnected islands and complexes of islands. This planning strategy could be included in policies for Green Energy in Europe and worldwide, such as the European Green Deal. More specifically, this paper proposed the integration of an HPP which follows the energy-efficient development and the new environmentally friendly behavior of insular power systems. The penetration level of both BESSs (8 MWh) and WTs exceeded the rate of 70%, achieving almost the maximization of RES penetration. At the same time, the operation of HPP achieved a reduction in fuel costs of more than 60%, making the power system more economically efficient, in addition to being more energy-efficient and environmentally friendly. Additionally, the fast response of the BESS could keep the system safe, from a stability perspective, during probable disturbances. As a novelty of this paper, the installation of an economically efficient HPP was proposed to contribute to the energy transition of island power systems, with the potential to also provide services for maintaining safety during unexpected events.

Author Contributions: Conceptualization, Y.K. and E.K.; methodology, Y.K. and K.F.; validation, K.F., Y.K., and E.K.; formal analysis, K.F and Y.K.; investigation, K.F.; resources, K.F., E.K., and Y.K.; data creation, Y.K. and E.K. writing—original draft preparation, K.F. and Y.K.; writing—review and editing, K.F., Y.K., and E.K.; visualization, K.F. and Y.K.; supervision, E.K and Y.K.; project administration, E.K.; funding acquisition, none. All authors read and agree to the published version of the manuscript.

Funding: This research received no external funding.

Conflicts of Interest: The authors declare no conflict of interest.

References

1. Panwar, N.L.; Kaushik, S.C.; Kothari, S. Role of renewable energy sources in environmental protection: A review. *Renew. Sustain. Energy Rev.* **2011**, *15*, 1513–1524. [CrossRef]

2. Akella, A.K.; Saini, R.P.; Sharma, M.P. Social, economical and environmental impacts of renewable energy systems. *Renew. Energy* **2009**, *34*, 390–396. [CrossRef]

3. "Clean Energy for EU Islands" Launched in Malta. Available online: https://ec.europa.eu/energy/news/clean-energy-eu-islands-launched-malta_en (accessed on 2 July 2020).

4. Smart Islands Initiative. Available online: http://www.smartislandsinitiative.eu/en/index.php (accessed on 2 July 2020).

5. Alves, M.; Segurado, R.; Costa, M. On the road to 100% renewable energy systems in isolated islands. *Energy* **2020**, *198*, 117321. [CrossRef]

6. Tsikalakis, A.G.; Hatziargyriou, N.D.; Katsigiannis, Y.A.; Georgilakis, P.S. Impact of wind power forecasting error bias on the economic operation of autonomous power systems. *Wind Energy* **2009**, *12*, 315–331. [CrossRef]

7. Papathanassiou, S.A.; Santjer, F. Power-quality measurements in an autonomous island grid with high wind penetration. *IEEE Trans. Power Deliv.* **2006**, *21*, 218–224. [CrossRef]

8. Beires, P.; Vasconcelos, M.H.; Moreira, C.L.; Peças Lopes, J.A. Stability of autonomous power systems with reversible hydro power plants: A study case for large scale renewables integration. *Electr. Power Syst. Res.* **2018**, *158*, 1–14. [CrossRef]

9. Vrettos, E.I.; Papathanassiou, S.A. Operating Policy and Optimal Sizing of a High Penetration RES-BESS System for Small Isolated Grids. *IEEE Trans. Energy Convers.* **2011**, *26*, 744–756. [CrossRef]

10. Zahedi, A. Maximizing solar PV energy penetration using energy storage technology. *Renew. Sustain. Energy Rev.* **2011**, *15*, 866–870. [CrossRef]

11. Zeng, P.P.; Wu, Z.; Zhang, X.-P.; Liang, C.; Zhang, Y. Model predictive control for energy storage systems in a network with high penetration of renewable energy and limited export capacity. In Proceedings of the 2014 Power Systems Computation Conference, Wrocław, Poland, 18–22 August 2014; pp. 1–7.

12. Solomon, A.A.; Kammen, D.M.; Callaway, D. The role of large-scale energy storage design and dispatch in the power grid: A study of very high grid penetration of variable renewable resources. *Appl. Energy* **2014**, *134*, 75–89. [CrossRef]

13. Hozouri, M.A.; Abbaspour, A.; Fotuhi-Firuzabad, M.; Moeini-Aghtaie, M. On the Use of Pumped Storage for Wind Energy Maximization in Transmission-Constrained Power Systems. *IEEE Trans. Power Syst.* **2015**, *30*, 1017–1025. [CrossRef]

14. Mercier, P.; Cherkaoui, R.; Oudalov, A. Optimizing a Battery Energy Storage System for Frequency Control Application in an Isolated Power System. *IEEE Trans. Power Syst.* **2009**, *24*, 1469–1477. [CrossRef]

15. Stroe, D.-I.; Knap, V.; Swierczynski, M.; Stroe, A.-I.; Teodorescu, R. Operation of a Grid-Connected Lithium-Ion Battery Energy Storage System for Primary Frequency Regulation: A Battery Lifetime Perspective. *IEEE Trans. Ind. Appl.* **2017**, *53*, 430–438. [CrossRef]

16. Wang, Y.; Tan, K.T.; Peng, X.Y.; So, P.L. Coordinated Control of Distributed Energy-Storage Systems for Voltage Regulation in Distribution Networks. *IEEE Trans. Power Deliv.* **2016**, *31*, 1132–1141. [CrossRef]

17. Zeraati, M.; Hamedani Golshan, M.E.; Guerrero, J.M. Distributed Control of Battery Energy Storage Systems for Voltage Regulation in Distribution Networks With High PV Penetration. *IEEE Trans. Smart Grid* **2018**, *9*, 3582–3593. [CrossRef]

18. Fiorentzis, K.; Karapidakis, E.; Tsikalakis, A. Cost Analysis of Demand-Side Generating Assets Contribution to Ancillary Services of Island Power Systems. *Inventions* **2020**, *5*, 34. [CrossRef]

19. HEDNO. Available online: https://www.deddie.gr/en (accessed on 30 July 2020).

20. Noninterconnected Island Systems: The Greek Case. *IEEE J. Mag.* **2017**, *5*. Available online: https://ieeexplore.ieee.org/abstract/document/7942257 (accessed on 3 July 2020). [CrossRef]

21. Karapidakis, E.; Georgilakis, P.; Tsikalakis, A.G.; Katsigiannis, Y.A.; Moschakis, M. Dynamic performance assessment of wind energy pump storage units in Crete's power system. *Mater. Sci. Forum* **2014**, *792*, 305–310. [CrossRef]

22. Fiorentzis, K.E.; Tsikalakis, A.G.; Katsigiannis, Y.A.; Karapidakis, E.S. Evaluating the effect of wind-hydro hybrid power stations on the operation of Cretan power system. In Proceedings of the 52nd International Universities Power Engineering Conference (UPEC), Heraklion, Crete, Greece, 28–31 August 2017; pp. 1–6.

23. Bouzounierakis, N.; Katsigiannis, Y.; Fiorentzis, K.; Karapidakis, E. Effect of Hybrid Power Station Installation in the Operation of Insular Power Systems. *Inventions* **2019**, *4*, 38. [CrossRef]

24. Katsigiannis, Y.A.; Karapidakis, E.S. Operation of wind-battery hybrid power stations in autonomous Greek islands. In Proceedings of the 2017 52nd International Universities Power Engineering Conference (UPEC), Heraklion, Crete, Greece, 28–31 August 2017; pp. 1–5.

25. Katsaprakakis, D.A.; Voumvoulakis, M. A hybrid power plant towards 100% energy autonomy for the island of Sifnos, Greece. Perspectives created from energy cooperatives. *Energy* **2018**, *161*, 680–698. [CrossRef]

26. Papaefthymiou, S.V.; Karamanou, E.G.; Papathanassiou, S.A.; Papadopoulos, M.P. A Wind-Hydro-Pumped Storage Station Leading to High RES Penetration in the Autonomous Island System of Ikaria. *IEEE Trans. Sustain. Energy* **2010**, *1*, 163–172. [CrossRef]

27. Psarros, G.N.; Papathanassiou, S.A. *Operation of a Wind-PV-Battery Hybrid Power Station in an Isolated Island Grid*; The Institution of Engineering and Technology: London, UK, 2018; pp. 4–6. [CrossRef]

28. Kazem, H.A.; Al-Badi, H.A.S.; Al Busaidi, A.S.; Chaichan, M.T. Optimum design and evaluation of hybrid solar/wind/diesel power system for Masirah Island. *Environ. Dev. Sustain.* **2017**, *19*, 1761–1778. [CrossRef]
29. Kennedy, N.; Miao, C.; Wu, Q.; Wang, Y.; Ji, J.; Roskilly, T. Optimal Hybrid Power System Using Renewables and Hydrogen for an Isolated Island in the UK. *Energy Procedia* **2017**, *105*, 1388–1393. [CrossRef]
30. Law 3468/2006 Generation of Electricity Using Renewable Energy Sources and High-Efficiency Cogeneration of Electricity and Heat and Miscellaneous Provisions. Available online: https://helapco.gr/ims/file/english/law_3468_2006_eng.pdf (accessed on 10 July 2020).
31. Regulatory Authority for Energy (RAE). Electrical System Operation Code for Non-Interconnected Islands (Version 2). April 2018. Available online: https://www.deddie.gr/el/themata-tou-diaxeiristi-mi-diasundedemenwn-nisiwn/ruthmistiko-plaisio-mdn/kwdikas-diaxeirisis-ilektrikwn-sustimatwn-mdn/kwdikas-diaxeirisis-mdn/ (accessed on 10 July 2020). (In Greek)
32. Hellenic Electricity Distribution Network Operator S.A. (HEDNO S.A.): Monthly Reports of RES & Thermal Units in the non-Interconnected Islands. Available online: https://www.deddie.gr/en/themata-tou-diaxeiristi-mi-diasundedemenwn-nisiwn/agora-mdn/stoixeia-ekkathariseon-kai-minaion-deltion-mdn/miniaia-deltia-ape-thermikis-paragogis/ (accessed on 30 July 2020).
33. Regulatory Authority for Energy: GIS Map. Available online: http://www.rae.gr/geo/?tab=viewport_maptab (accessed on 30 July 2020). (In Greek)
34. Siemens-Gamesa: SG 2.6-114—Boosting Production at Sites with Medium and High Winds. Available online: https://www.siemensgamesa.com/-/media/siemensgamesa/downloads/en/products-and-services/onshore/brochures/siemens-gamesa-onshore-wind-turbine-sg-2-6-114-en.pdf (accessed on 30 July 2020).
35. Katsigiannis, Y.A.; Stavrakakis, G.S. Estimation of wind energy production in various sites in Australia for different wind turbine classes: A comparative technical and economic assessment. *Renew. Energy* **2014**, *67*, 230–236. [CrossRef]
36. JRC Photovoltaic Geographical Information System (PVGIS)—European Commission. Available online: https://re.jrc.ec.europa.eu/pvg_tools/en/tools.html#PVP (accessed on 30 July 2020).
37. Manwell, J.F.; McGowan, J.G.; Rogers, A.L. *Wind Energy Explained: Theory, Design and Application*; John Wiley & Sons: Hoboken, NJ, USA, 2010.

Publisher's Note: MDPI stays neutral with regard to jurisdictional claims in published maps and institutional affiliations.

Review

A Review of Energy Storage Participation for Ancillary Services in a Microgrid Environment

G V Brahmendra Kumar and K Palanisamy *

School of Electrical Engineering, Vellore Institute of Technology, Vellore 632014, India;
brahmendrakumar.g@gmail.com
* Correspondence: kpalanisamy@vit.ac.in

Received: 14 October 2020; Accepted: 14 December 2020; Published: 16 December 2020

Abstract: This paper reviews the energy storage participation for ancillary services in a microgrid (MG) system. The MG is used as a basic empowering solution to combine renewable generators and storage systems distributed to assist several demands proficiently. However, because of unforeseen and sporadic features of renewable energy, innovative tasks rise for the consistent process of MGs. Power management in MGs that contain renewable energy sources (RES) can be improved by energy storage. The energy storage systems (ESSs) have several merits, such as supply and demand balancing, smoothing of RES power generation, enhancing power quality and reliability, and facilitating the ancillary services like voltage and frequency regulation in MG operation. The integration of ESS technology has become a solution to the challenges the power distribution networks face in achieving improved performance. By simplifying a smooth and robust energy balance within the MG, storage devices match energy generation to consumption. MG, and its multidisciplinary portrait of current MG drivers, tasks, real-world applications, and upcoming views are elucidated in this paper.

Keywords: microgrid; ancillary services; renewable energy sources; energy storage; power management

---

1. Introduction

As a result of improper regulation followed in the electric power industry and the continuing problems in the environment along with increasing energy consumption have led to an increase in installed capacity of distributed generation (DG) sources and energy storage systems (ESSs). These sources include various technologies in combined heat and power operation or purely for electricity production like microturbines, diesel engines, and fuel cells as well as photovoltaic (PV), hydro turbine, small wind turbines, etc. To balance the load and power of renewable sources, the stored energy is controlled, over time domain, lowering the full charge of energy at this level of mutual coupling [1]. Presently, the smart grid is the leading idea in the electric power industry. The key objective of developing a smart grid is to provide consistent, high-quality electrical power to digital societies in environmentally friendly and sustainable ways. The advanced structures that can simplify the links of many AC and DC generation systems, energy storage operations, and several AC and DC loads with the optimal asset utilization and operational efficiency are the most significant features of the smart grid. To achieve these objectives, power electronic technology plays a substantial role in interfacing various sources and loads to a smart grid [2].

Recently, microgrid (MG) technology has received significant attention from utilities around the world as a key approach to cost-saving asset replacement and strengthening of networks [3]. Electrical and energy engineers are now faced with a new scenario in which the grid must be integrated with small, distributed power generators and distributed storage (DS) devices. The new power grid, also known as the smart grid, would deliver electricity from producers to customers using digital technologies to monitor household appliances to save energy, reduce costs, and increase efficiency and transparency. In this context, the energy system must be more accessible, smart, and distributed. Without the use of DSs to deal with energy balances, there is no point in using DG. MG networks, also referred to as mini-grids, are becoming a key concept for the integration of DG and DS systems. The model is designed to cope with the penetration of RESs (Renewable Energy Sources) that can be practical if the end user is able to produce, store, monitor, and manage part of the electricity consumed. The end user is not just a consumer, but part of the grid with this paradigm change [4]. Using current technical progress, it is stress free and probable to have battery energy storage systems (BESSs) and loads and MG systems with the aggregation of localized energy sources [5]. MG technology has progressed over the years and is becoming more and more advanced. In addition to communication technology, electrical energy distribution and generation have been reformed with modern progression. Through an improved aggregation of RESs, the MG is composed to turn into a fundamental feature of new power systems. This innovative era technology is to assist service providers and consumers to yield absolute control over the cost of energy (COE), system reliability, and energy sustainability. MG technology is making rapid progress toward excellence with a substantial contribution from all stakeholders.

Various MG definitions [6] and their efficient categorization methods [7] can be seen in the literature. The MG Exchange Group from the U.S. Department of Energy and the Ad Hoc Group of research and development experts define the MG as follows: "A group of interconnected loads and DERs (Distributed Energy Resources) acting as a single controller to the grid within specified electrical limits. The MG can be connected and disconnected from the network so that it can operate in both grid-connected and island mode [8]".

A typical MG system is presented in Figure 1. This definition contains three requirements: (1) The part of the distribution system comprising a MG can be identified as different from the rest of the system; (2) resources associated to the MG are controlled by each other rather than by remote resources; and (3) whether it is connected to or not the larger grid, the MG can work [9].

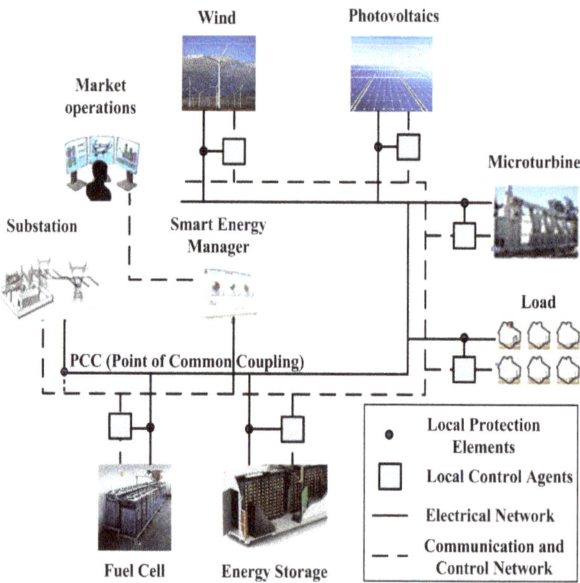

Figure 1. A typical system of microgrid (MG).

MGs are more effective in owning and managing local problem areas and can also be used as a possible tool to achieve a "self-healing" smart grid for the future [10]. Societies can develop grid architecture models as smart supergrids or virtual power stations that do not have a local generation balance, load balancing, or isolating segments of the grid, which are more compelling designs.

Smart supergrids can detect, isolate, and restore capabilities to reduce congestion, routing energy around failures, and reduce recovery time from failures. Software and analytics can be used to manage a wide range of DERs of virtual power plants, although networked MGs can also serve as virtual power plants [11]. It is important that MGs become the dominant strategy for utilizing large amounts of intermittent renewable energy compared to alternative smart grid paradigms and that the benefits in terms of cost are sufficient.

One focus area is the voltage control market in MG distribution networks [12]. Some scientists suggest that in the future each MG will serve as a virtual power station (i.e., single aggregated DER) and each MG bids energy and ancillary services to the external power system based on the aggregation of offers from DERs within the MG. They design a day-ahead market for reactive power for the power flow from large generators to customers over a radial transmission and distribution network and provide a mechanism for optimal market settlement. This is a very similar vision as the roadmap in New York that proposes opening up the wholesale electricity market to DER aggregators [13].

Regarding the distribution grid's ancillary services, numerous studies have been carried out on MGs and grid stability relationships. MG with ESSs can perform additional tasks such as maintaining the local voltage at a given level, providing a backup system for critical local loads, etc. [14]. In [15], the optimum management of DER and energy storage can benefit the grid power balance. Recent research has proposed a state-of-the-art technology called transactive energy to operational resource planning to minimize energy costs and improve the delivery system stability through demand or price-tracking mechanisms [16]. The use of ESS in distribution networks is a good option to mitigate power system problems from the generation and transmission networks' operation to the small-scale distribution network and microgrid applications [17]. ESSs' facilities are numerous and are expected to progress in the future.

Microgrids are relatively small, controllable power systems consisting of one or more units of generation linked to near users, which can be operated with or without the local bulk transmission system. They can also use energy storage such as batteries in electric vehicles to balance microgrid output. Microgrids can contribute to deploying energy from zero emissions, minimizing energy loss through transmission lines, controlling power supply, and increasing grid resilience to extreme conditions [18].

Innovative business models like power purchase or energy-sale contracts and the development of properly owned operations play a major role in MG scalability. Power purchase agreements (PPAs) are set to play a bigger role in the MG market when MG design and procurement is simplified [9]. The PPA is currently a very successful business model in the U.S. residential and commercial solar PVs market because they can collect tax and other related incentives while avoiding the large initial cost of capital for a plant that hosts the system.

The PPA infrastructure is owned and rented by a third party to provide electrical and related services to end users. In the MGs' sector, improved reliability, sustainability, and economic benefits such as energy cost savings can be placed on the market. In combined projects, such as cooling, heating, and power, thermal energy can also be combined with electricity in the PPA. Operations and maintenance can be reasonably expected to be part of PPA, since the revenues from the PPA depend on the operating systems to their full potential.

In the late 1990s, researchers and engineers in the U.S. and Europe began looking for decentralized solutions that could manage the integration of tens or thousands of DERs to enhance reliability and resilience against natural disasters, physical and cyber assaults, and power failures [19]. The solution to this is a network architecture that can manage power generation and demand in subdivisions of the grid, isolated from an automated large grid to provide critical services even when the grid fails.

MG development in the U.S. was driven mainly by their ability to increase the resilience of critical facilities such as transportation, communication, and emergency response infrastructure, which can quickly bounce back from complication and reliability (at times acceptable levels of services) [20]. One of the main regions of the U.S., the Northeast, has suffered billions of dollars lost in aging infrastructure and often severe weather events in recent years. Thus, states are exploring the possibilities of expanding MG services in critical infrastructure to serve entire communities.

New York State's "New York Prize", a competition of $40 million to help communities deal with issues ranging from feasibility studies to implementation, is the most notable example of state support for community MGs [21]. In the U.S., states seek MGs as an alternative to retiring power generation and to reduce congestion points in the transmission and distribution system. Climate scientists have concluded that human societies should reduce the share of electricity from the burning of fossil fuels from 70% (in 2010) to below 20% by 2050 to prevent global average temperature rise above preindustrial levels, now accepted as the threshold for "safe" and "dangerous" climate change [22].

To address these gaps, many energy sources are decentralized, intermittent, and uninterrupted, and it is challenging for them to integrate the existing grid designed to supply one-way power flow from centralized power stations to customer loads. Deploying intermittent renewable energy in MGs with flexible loads and storage technology enables local supply and demand balance, making it possible for widespread renewable distribution. Installation for distribution service providers has potential to change the way a small source or electricity user benefits the main grid rather than tracking and coordinating the net loading profile of thousands or millions of individual DERs [23]. A wide range of criteria, including carbon emissions, investment costs, electricity costs, and others, are currently being investigated for the design of MGs using sophisticated analytically developed approaches.

A wide variety of DERs can be used in MGs in various multidisciplinary studies, reviewed in [19]. MGs appear to be technology agnostic, and design choices depend on specific project requirements and economic considerations. The inclusion of ESSs prevents MG deficiencies. Since most MG sources lack the inertia of large synchronous generators (SGs), a buffer is needed to minimize generation and demand imbalances. The load diversity of larger geographical areas is also lacking in MG systems, which means they have to deal with much greater relative variability. There are a number of currently emerging ESS technologies that could play a possible role in MGs [24]. The ability of ESS facilities to provide ancillary services, such as voltage control support, spinning reserves, load following, and peak shaving among others, has been briefly discussed in the following sections.

In this paper, Section 2 describes the generation and storage options. Ancillary services and their classifications and available ancillary services across the globe are explained in Section 3. The drivers of MG development and deployment and applications of MG are studied and reviewed in Section 4, and Section 5 is the conclusion.

2. Generation and Storage Options

To improve the reliability and power quality of the grid, storage systems along with RES are used. Storage features support the needs of a grid network [25] and:

- Ensure the grid energy balance,
- Provide fault ride-through (FRT) capability under dynamic variations, and
- In MGs, assist the smooth transition from islanded mode to normal modes.

Some examples of options available for today's generation and storage, including their advantages and disadvantages, are listed in Table 1 [26–32] below. Fuel cells (FCs), batteries, flywheels (FES), and supercapacitors (SCs) are the most commonly used energy storage devices [25]. Among these, flywheel is not currently the most attractive solution because of its self-discharge problems, the need for vacuum chambers, and the maintenance of superconducting bearings [33].

Table 1. Overview concerning the generation and storage options in MGs (Microgrid).

Reference	Category	Storage Options Employed	Benefits	Drawbacks
[26]	Generation	Diesel and Spark Ignition (SI) reciprocating internal combustion engines	Easily dispatchable in nature. Faster start-up and load-following. Used for combined heat and power (CHP).	Particulate and Nitrogen oxide emissions. Likely emission of greenhouse gases. Generation of noise.
[27]		Microturbines	Dispatchable. Multiple fuel options. A lower degree of emissions. Simplicity under mechanical aspects. CHP-capable.	The maintenance cost is high. Cooling is necessary, even if heat retrieved is not reusable.
[28,29]		FCs (including molten-carbonate, solid oxide, alkaline, and phosphoric acid, low-temperature PEM)	Dispatchable. Zero on-site pollution. CHP-capable. Greater efficiency available versus micro turbines.	Comparatively, they are expensive. Limitations of mechanical strength and fatigue. It is less mature than chemical batteries. The current cost is too high to make them commercially competitive.
[27,29]		Renewable generation (solar PV cells, small wind turbines, and mini-hydro)	Cost effective in terms of fuel generation. Zero emissions. Maintenance requirements are lower than traditional fuel sources.	The upfront cost is higher. Geographical constraints. Need a capable load-following generator. Lack the much-needed efficiency. Variable and regarded as uncontrollable in nature.
[30,31]	Storage	Batteries (including lead-acid, sodium-sulfur, lithium-ion, and nickel-cadmium)	A long history of R & D. Round-trip efficiency is between 75–90%. High performance and lower maintenance.	A limited number of charge–discharge cycles. Complications in terms of waste discharge. Battery degradation costs.
[30]		Flow batteries (FBs) referred to as regenerative FCs (Comprised Zn-Br, polysulphide bromide, vanadium redox)	Decouple power and energy storage. Round-trip efficiency is up to 75%. Ability to support continuous operation under maximum load. Total discharge is possible without any risk of damage.	Relatively under the early stage in terms of deployment. Lower power density. More complex. Components and chemicals used in the flow batteries are still comparably expensive.
[32]		Hydrogen from hydrolysis	Clean. Can store for a long period.	Relatively low end-to-end efficiency. Challenges concerned with hydrogen storage. Components' cost is high.
[32]		Kinetic energy storage (flywheels)	Fast response. Overall costs are low. High in terms of charge–discharge cycles. Round-trip efficiency is 85%.	Discharge time is limited. High standing losses. Maintenance is required.

Table 1. *Cont.*

Reference	Category	Storage Options Employed	Benefits	Drawbacks
[34]		Pumped Hydro Energy Storage (PHES)	Free from environmental impacts. Sources are plentiful, clear, and reliable. No reserve shortfalls. Comparatively economical. Very long lifetime. Round-trip efficiency is 70–80% based on the distance and gradient between upper and lower reservoirs.	Expensive to build. Geographical constraints. Construction period is longer. Maintenance is required. Uncertainty of ease of use of water; if the water is not available, difficulty in producing the electricity. Overflow impacts.
[34]		Compressed Air Energy Storage (CAES)	Energy storage capacity is high Cost/kWh is low. Long lifetime. The need for power electronic converters is less.	The necessity for fuel and underground cavities. Investment cost is high. Geographical constraints. Efficiency is low.
[34]		SC	High power and energy density compared to normal capacitors. Highest round-trip efficiency up to 96%. Speed charging ability and faster response time. Environmentally friendly.	The self-discharge rate is high and low energy density compared to batteries. It cannot be utilized in AC and high-level frequency circuits.
[34]		Superconducting Magnetic Energy Storage (SMES)	Power capability is high. 95% round-trip efficiency. No environmental impacts. Faster response time. Capable of part and deep discharges.	Lower energy density. Raw materials, operation, and manufacturing processes are expensive.

The FC changes the chemical energy in hydrogen to electrical energy. Modern heat engines use intermediate mechanical energy conversion to generate electricity from chemical energy, resulting in lower efficiency compared to FCs. FCs combine engines and batteries to their best advantage. As long as fuel is available, they act as engines and there is no intermediate mechanical energy conversion, and the FC characteristics are similar to batteries under load conditions. In relation to other challenging technologies, FCs have an extended life span of 20 years. However, FC gets great prices, especially for electrolyzers and efficiency issues. The main drawback of FC is that it is not subject to fast load variations, which lead to fuel starvation in the cells, which is short of cell lifetime. This describes the energy storage mechanism in the form of hydrogen gas in FC [35]. Compared to SCs' high discharge currents and low power densities, batteries have good energy densities as storage elements but suffer from longer recharging time. The SC storage element, which has a higher power density, has low energy density [34].

3. Services in Electric Power Industry

Services involved in the electric power industry are as follows.

3.1. System Services

System services are all services given by some system function (for instance, a system operator or a grid/network operator) to the users of the system. Power transmission, generation, and energy supply are basic system services [36].

3.2. Ancillary Services

Some users, such as generators, use the system frequency or voltage at a time when they are connected to the system to support these systems. These services provided by consumers are referred to as ancillary services because they are associated with energy production or consumption. These services are vital to maintaining the stability and reliability of the operation of the electrical system. From the traditional point of view, ancillary services are provided by large power plants and equipment with adequate capacity and capability. However, with improvements in MG technology, researchers and industries have been re-examining the role of MGs in promoting services. MGs provide ancillary services as long as they meet the technical requirements, and it is believed that ancillary service provision is one of the most important benefits of providing MGs.

From the perspective of aggregation mechanisms along with control capabilities on distributed power units, the technical feasibility of providing frequency and voltage control services through MG has been examined [30]. DGs can support voltage and frequency stability of power systems with appropriate coordination. Most DERs are interconnected by power electronic devices, and researchers have found that power electronic interfaces can provide some support for DER integration, particularly reactive power-related services. Reference [27] demonstrated the ability to use DGs with power electronic interfaces to provide ancillary services such as voltage control.

The possibility of supporting voltage through wind turbine generators has been investigated [26]. In [37], two models were developed based on the MG central controller concept, which allows the MG to participate in both energy and ancillary services' markets. The capacity underlying single MG might be too small in provisioning ancillary services and investigated the underlying technical problem of frequency control by multiple MGs and proposed a centralized control approach and decentralized approach for aggregation. The generating capacity of DGs allows MGs to benefit from both power generation and active power reserve. For profit maximization, optimizing the bidding strategy for both the energy and ancillary service markets to simultaneously participate has been identified, and this is a problem that has attracted the attention of researchers [29].

Different types of DGs may have different behaviors in providing ancillary services due to their distinct fundamental characteristics. Therefore, the profitability for those DGs should be investigated separately. The study involving [27–29] gives detailed modeling of how profit can be achieved by providing active power reserve and comparing the profitability of active power storage from different energy sources.

The author argues that renewable sources are not profitable for active power reserve, as they have a high opportunity cost caused by their low operating cost (e.g., wind energy generators and solar energy generators have zero fuel cost due to the free wind and sunshine, which makes their operating cost lower than fuel-based generators). In addition to DGs, load and ESS are also good candidates for frequency control services. Loads that participate in frequency control service must be dispatched and interruptible; also, the loads prefer to be used as a contingency reserve (e.g., a spinning reserve), which is often not used [37]. Some research has been done to minimize the manipulation of the load while providing frequency control service in [38].

Loads have several advantages over the generators on providing ancillary services. Compared to generators, loads are usually smaller in size, so an aggregation of the small load is more reliable than a single, large generator. Secondly, loads can react faster since curtailment of the load is usually faster than ramping up a power plant. Thus, they are well matched to the fast, short, and infrequent events. Finally, using loads can prevent additional investment in generators and transmission lines. In order to be good candidates for spinning reserve, loads must have the following characteristics: capability of storing energy, control capability, fast communication, adequate aggregation size, low cost, quick response, and restoration [32,39]. The above features are also noted in [40], in which the authors demonstrated that the thermostatically controlled loads that can be quickly disconnected are ideal for providing a frequency control service because they can be energy efficient and cost effective. In [41,42], the capability potential of thermostatically controlled load for ancillary services is further illustrated. The efficiency of the load is taken into account not only from a technical perspective but also from market and policy barriers. In [43], an optimization approach to increase the profits of the load in the ancillary service market was proposed.

System and ancillary services are interdependent and complex because both system and ancillary services are required simultaneously by the same supplier [44]. Applications of ESS provide a variety of services. In this, the terms and conditions of the specified 'service' and 'application' are reviewed and interchanged. The term 'service' refers to the electrical operation accomplished by ESS with its power conversion system, and the term 'application' defines connection and the location of the grid and the ESS functionality concerning its infrastructure as well as its technical features. An overview of ESS pilot projects around the world for various application areas is presented in Table 2.

3.3. Classification of Ancillary Services

The system operator must ensure the required level of quality and safety to maintain the integrity and reliability of the system, take preventive measures for contingency control and perform many other duties. The operator should be able to adjust the frequency and voltage of the system within certain limits, maintain the reliability of the system, avoid overload on the transmission system and, if necessary, re-establish the system [46].

In order to maintain system reliability in view of events and contingencies, a suggested list in Table 3 for the system operator is presented. It has been concluded that all ancillary services are important and they meet the previous requirements. The detailed classification of ancillary services is presented in Figure 2. Ancillary services are fretted through the delivery of power, trade, and dispatch. They are usually characterized by the benefits they provide to market members [47]. The following Table 3 lists the different classifications of ancillary services in the literature from various sources.

Table 2. Overview of ESS (Energy Storage System) pilot projects across the world for various application areas [45].

ESS Facility	Projects	Capacity	Application Area
FES	Beacon power company Boeing Phantom Works Piller power system Ltd.	20 MW/5 MWh plant 100 kW/5 kWh 2.4 MW	Frequency regulation, voltage support, and power quality. Power quality, peak shaving. FRT capability, backup power.
BES	BEWAG, Berlin PREPA, Puerto Rico Chino, California Abu Dhabi Island, UAE PacifiCorp VRB facility, Utah, U.S. SEI VRB ESS facility, Japan	8.5 MW/8.5 MWh 20 MW/14 MWh 10 MW/40 MWh 40 MW 250 kW/2 MWh 1.5 MW/3 MWh 500 kW/5 MWh	Spinning reserve, frequency control. Spinning reserve, load levelling. Load levelling. Voltage support, load shifting. Power quality. Voltage support, peak shaving.
SC	NEC, Japan Siemens, Germany	3.5–12 V, 0.01–6.5 F 5.7 Wh, 2600 F	Power quality. Smoothing power output.
SMES	Nosoo power station, Japan Upper Wisconsin, USA Chubu Electric Power Co. (Company), Japan	10 MW 3 MW/0.83 kWh 5 MW	Power quality, system stability. Reactive power support. Voltage support.
FC	FC Power Plant, California Naval Air Warfare Center, California Ongoing projects: IdealHy, Netherlands; Sapphire, Norway; RE4CELL, Spain; SmartCat, France	2.8 MW 5 kW	DG, electric utility. Power quality, backup power, and small DGs.
TES	Highview Power Storage Co., UK Torresol Energy, Spain	300 kW/2.5 MWh 15 MW	Load Shifting, managing DG and DS with large-scale penetration.
CAES	LAES pilot plant, Birmingham, Advanced adiabatic-CAES plant, China Supercritical-CAES	350 kW/2.5 MWh 10 MW 1.5 MW	Frequency and voltage control, peak shaving, load shifting, and intermittent RES.
PHES	Rochy river PHS plant, US Okinanawa Yanbaru plant, Japan Ikaria Island HPS, Greece	32 MW ~30 MW 2.655 MW	EMS in fields of time shifting, supply reserve, frequency control, and nonspinning reserve.

Table 3. Key characterizations of ancillary services.

References	Source	Contribution
[46–52]	CIGRE/FERC/Power System Economics and other Authors	Frequency and voltage control services Black start Scheduling and dispatch Financial trade enforcement Transmission security System security Load-Following Loss Compensation Energy imbalance Operating reserve Reactive power control Real-power balancing

The different lists of ancillary services mentioned above are mainly about how they divide and integrate various types of services. Broadly, each of these services can be classified as one of the following three main categories:

- Frequency control services,
- Network control services, or
- System restart services.

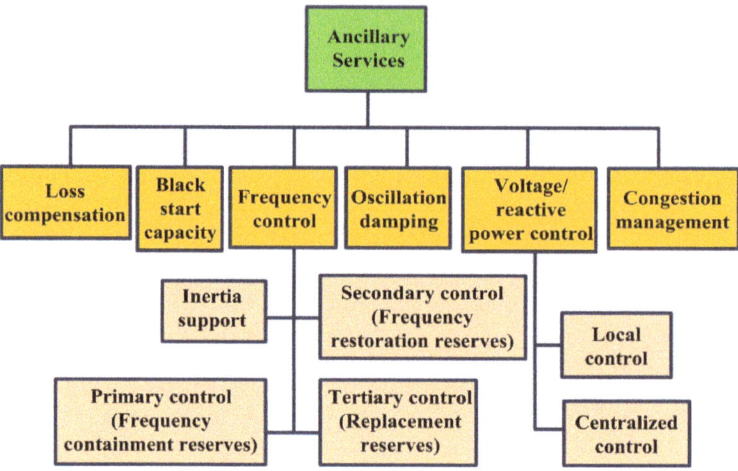

Figure 2. Classification of ancillary services.

3.3.1. Frequency Control Services

Frequency is the measure of the active power produced and consumed, both of which must be balanced to allow the operation of an AC system. The basic functionality of the power system is represented in Figure 3.

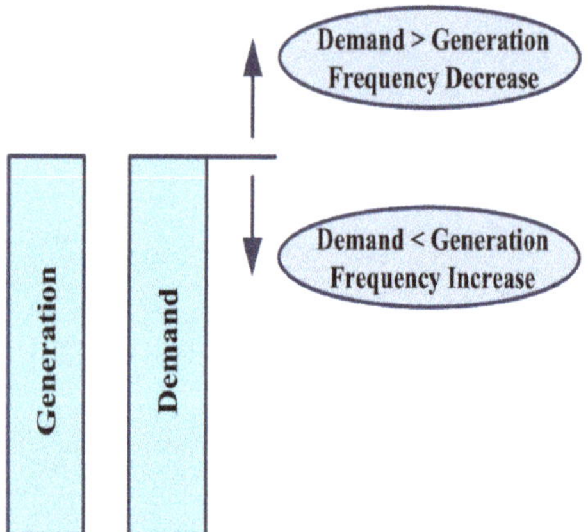

Figure 3. The basic behavior of the power system network.

Therefore, the frequency is considered as a pointer to standardize the active power output to balance them out [51]. For system security, frequency control is required. The variation in frequency is usually within a certain acceptable range for the secure operation of the electrical system and the safe and reliable operation of connected devices. Since power consumption is different, it is essential to control the active power output accordingly. The following Figure 4 shows the process at various frequency levels [52].

Technical factors underly frequency control services. Deployment times are important, specific characteristics of frequency-related ancillary services. The 'Deployment Start' is the entire period between the request of a particular network operator and the response of the service provider [53]. 'Deployment End' means the maximum time the service can be delivered from demand.

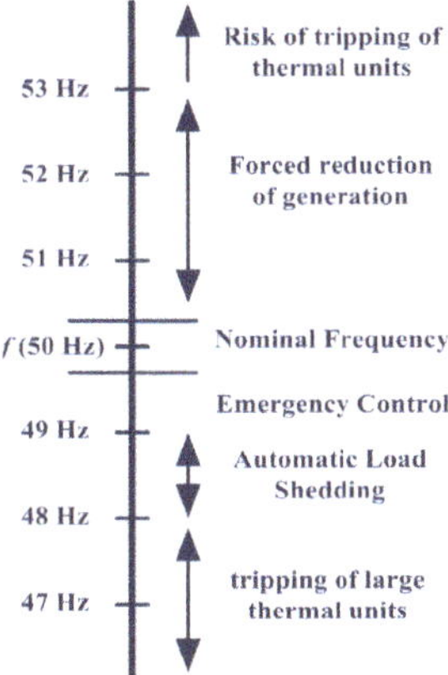

Figure 4. Different frequency levels of operation.

Levels of Frequency Control

To maintain the balance between generation and demand, the following levels of controllers are used [54], as shown in Figure 5:

(1)   Primary,
(2)   Secondary,
(3)   Tertiary,
(4)   Time control.

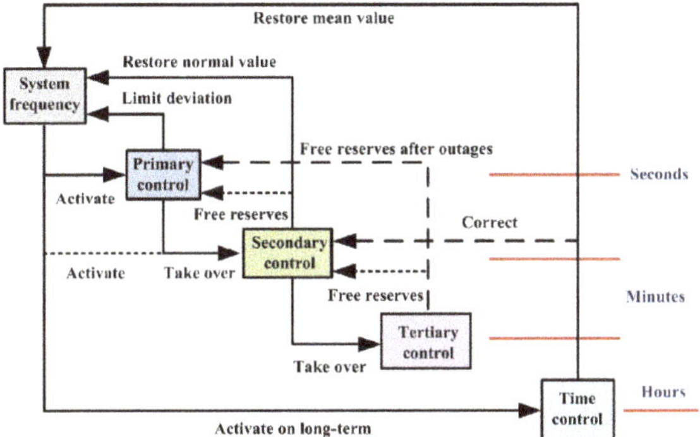

Figure 5. The function of various types of frequency reserve controls in the UCTE (Union for the Coordination of Transmission of Electricity) region [54].

All control levels are different with the change in response to time and the approach used to understand the basic operational perspective [55].

- Primary control is initiated within seconds as a collective action by all concerned parties or transmission system operators (TSOs).
- Secondary control replaces the primary control over minutes and is enforced by the responsible parties/TSOs.
- Tertiary control partly completes the secondary control and then replaces it with rescheduling generation and is enforced by responsible parties/TSOs.

- Time control corrects the global synchronous time deviations as a joint action by all parties on a long-term basis.

i.   *Frequency Control for Primary*

The primary control is the active power output of the generating unit to control the variation in frequency and control the use in controllable loads. It is specially designed to control the frequency during major generation and load interruptions. Therefore, it is essential for the stability of the power system. This control is automatically done by all the generators in the synchronous area equipped with the speed governor. The self-regulation effect of frequency-sensitive loads such as induction motors or the behavior of frequency-sensitive relays that connect or disconnect certain loads at specified frequency limits are also involved in this control [56].

Primary control source: To reduce unforeseen transits after a significant disturbance, primary frequency reserves should be uniformly distributed over the network. In the case of islanding, the uniform distribution also helps to maintain the stability of the system. Primary control and load monitoring are particularly suited for hydraulic turbines/stations. In these machines, the unit load can be easily adjusted and the units can be installed/completed without significant stress or loss of output. In some cases, further consideration of water release (irrigation, minimum flow, etc.) should be restricted to the load following. If there is no free board at canal-based hydro stations, no primary control or load following is possible [57].

Gas turbine stations are suitable for load following because they are controlled in/out but do not work for primary control, as they must be operated at a constant firing temperature (equal to efficiency and life) [52,53]. Increasing the fuel injection to increase the megawatts will cause the firing temperature to be higher than it can sustain continuously. Reducing the fuel injection to decrease the MW, thus reducing the firing temperature and changes in fast/frequent firing temperature, causes thermal over-stressing. By lowering the firing temperature, the efficiency is also reduced. For steam turbines, efficiency decreases slightly at partial load because the steam parameters can be held at rated levels and can also provide primary control and load monitoring. In a short period, the active power from the steam turbine can be improved by the action of the valves providing the primary response. For longer periods, it is important to adjust the primary input power such as the fuel flow [58]. When two systems are interconnected, High-voltage DC transmission can also be used as a primary controller, particularly in installed power systems where the frequency is closely controlled and enough spinning reserves are available.

ii.   *Frequency Control for Secondary*

Secondary frequency control is a centralized automatic control that optimizes the generating units' active power production to restore their target values after frequency mismatch and interchange with other systems. This process can be done by adjusting the generator set point or reference point or by starting and stopping of power plants. Secondary control should only be used for the generating units placed in the control area where an imbalance has occurred [44]. To reduce the area control error (ACE) is the goal of secondary frequency control. The primary frequency control must be adjusted to maintain the active power balance after an abrupt change in load capacity. This will result in a frequency change in the power plants due to a permanent drop of the primary frequency control. However, this can lead to the variation of power transfer between the control areas from the intended power transfer in the control system. Automatic secondary frequency control restores the scheduled power transfer.

Secondary frequency control is used to bring the ACE to zero using a proportional-integral (PI) controller and filters. Secondary frequency control in UCTE (Union for the Coordination of Transmission of Electricity) is also known as load frequency control (LFC) and is called automatic generation control (AGC) in North America.

iii.   *Frequency Control for Tertiary*

Tertiary frequency control applies to manual changes within the dispatching and responsibility of generating units. The control is used to recover primary and secondary frequency control reserves, manage transmission network congestion, and restore frequency or interchange their target value when this last function cannot be performed by the secondary control [44].

iv.   *Time Control*

Time control makes the average frequency equal to the normal frequency of 50 Hz. If the average frequency change reaches the above threshold point, the fixed frequency point in the entire synchronous zone is either set at 49.99 Hz or 50.01 Hz for complete, one-day periods [54]. The realization of these controlling frequency stages for a generating unit with the support of feedback control loops is presented in Figure 6.

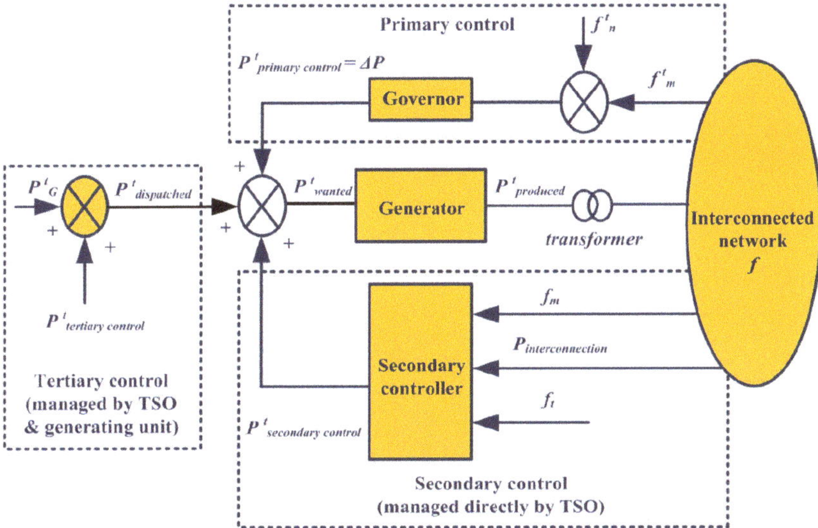

Figure 6. Three operational frequency regulations of a generator.

Frequency Reserves

Frequency reserves are necessary to maintain the integrity of the system in case of active power consumption and irregularities in production. These conditions may be caused by generation failures or variations in load. The required amount of reserves depends on the probability of multiple generation failures and variations in load. For example, if a system is loaded with more arc furnace, the reserve requirement of the system increases because the arc furnace load is very unpredictable. However, if a generator is subject to frequent interruptions, the required reserves will not change but the use of reserves will increase. The policy difference between UCTE and NERC (North American Electric Reliability Corporation) and their terminology are presented in Table 4.

The types of frequency controls listed above in the review have some reserves in order to address variations in frequency as a response to the generation capacity/load management [60]. Figure 7 shows the use of reserves for various types of a frequency control for a typical generating unit. Frequency reserves can be divided into the following broad groups:

i. Spinning Reserves/Reliability Reserves

The spinning reserve is the additional generating power required to improve the power output of the generators already associated with the power system. Spinning reserves are necessary as demand may change and fast responses are required according to short-term standards. This category includes very quick operation units or controllable loads [59]. Such reserves are unable to continue for long durations at increased power consumption. Therefore, they will inevitably be supplemented by additional reserves. Coordinated internal combustion engines or hydro units can provide this.

ii. Supplementary Reserves

Their response time does not have to be as fast as spinning reserves, but they must be able to operate for longer duration at increased power output. For activation, they often require manual intervention. The reserve is also provided by the generation of hot standby units [59].

iii. Backup Reserves

These are reserves that can last for a considerable period of time (within hours) but should not come online for a certain amount of time (usually 30 min or more) [59].

Table 4. Policy differences between UCTE and NERC (North American Electric Reliability Corporation) [59].

Reserves	UCTE	NERC
Terminology	Primary control reserve Secondary control reserve Tertiary control reserve	Frequency responsive reserve Regulating reserve Spinning reserve Non-spinning reserve Non-spinning reserve Supplemental reserve
Regulating	UCTE recommends a secondary reserve control requirement based on the statistical equation and mainly based on load variability. However, both contingencies and normal variations are subject to secondary reserves. Compliance measures are not available.	CPS enforcement provisions are imposed by the NERC but do not have a regulation on the amount of the current reserve regulating requirements. The requirements of the CPS are based mainly on the time of day and season.
Following	No UCTE requirements. Used to minimize ACE for slower normal variations in a control area.	NERC does not provide any standard or direction.
Replacement/ Contingency	The DCS criterion is identical. Return ACE in 15 min to zero. Sufficient of these reserves should be provided to support the most significant contingency.	DCS would return ACE to zero or its pre-disruption point in 15 min, if negative. Sufficient contingency reserves needed to recover the largest contingency. For many regions, at least 50 percent of the spin is required.
Primary	Complete response at 200 mHz. Characteristics of response based on UFLS relay setting and safety margin of 200 mHz Peak insensitivity of 20 mHz.	Only a requirement for frequency bias as a part of 1% peak ACE calculation. The dead bands of governors usually settled at 36 mHz and dropped by 5%.
Ramping	No UCTE requirement for the ramping reserve.	No constraints. Used for rare severe events that do not take place immediately.
Secondary	The UCTE policy recommends that the secondary reserve be initiated within a maximum of 30 s after the disturbance and returned to the initial ACE within a maximum of 15 min.	The Contingency reserve and the Ramping reserve are used as a secondary reserve to restore the frequency to its nominal value and to reduce the ACE back to zero.
Tertiary	The need for tertiary control reserves is greater than the largest contingency. It is not necessary to replace reserves as long as possible.	No quantifiable requirement, but the contingency reserve has replaced within 105 min of contingency.

CPS, control performance standard; UFLS, under frequency load shedding; DCS, distribution control standard.

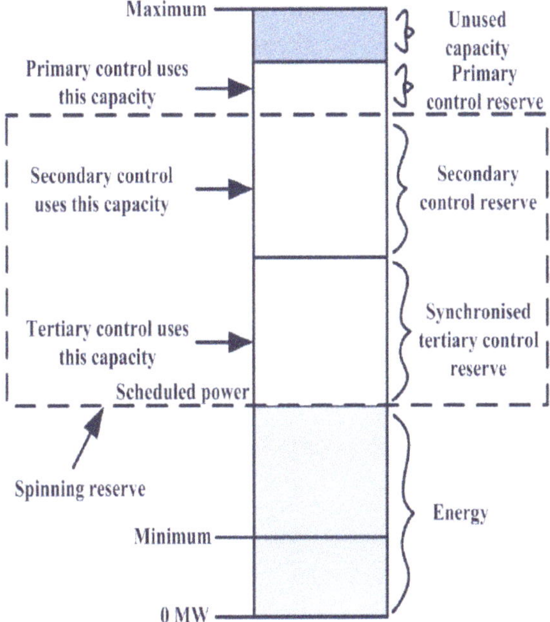

Figure 7. Control levels for a generating unit.

3.3.2. Ancillary Services for Voltage Control

The additional standard parameter for power quality is system voltage. The voltage is regulated by controlling the reactive power injection and the drawl within the power system. Network flows create voltage rises and drops as a result of interactions between flows and the transformer line inductance and capacitance.

Requirements for Voltage Control

The system voltage control is used to sustain the voltages at several nodes in the system, which requires precise bounds and the reactive power required in the system. Due to high inductance lines and transformers, the reactive power is not transmitted well through the grid, thereby supporting the reactive power as much as possible. Voltage control is necessary for the following reasons [61]:

- The equipment of voltage supply should be in its design bounds for safe process and excessive implementation.
- The system voltage varies then creates the changes in reactive power that widely affect the system losses.
- Voltages may also limit the system's transfer capability.
- Reactive power injection and absorption are also important for maintaining system stability, especially to avoid contingencies, which can lead to voltage collapse. Reactive power must have sufficient capacity to meet the required demands and the margin of reserve for possible outcomes. Local voltage regulation is a consumer service designed to meet consumer reactive power requirements and monitor each consumer's impact on network voltage and system failure. Therefore, power factor problems at a customer site do not affect power quality elsewhere in the grid.

Stages of Voltage Control

The overall voltage regulation function can be organized into a hierarchy of three levels.

- Primary voltage control can be local automatic control, which saves the voltage at the generating bus at a set of points. The task is fulfilled by an automatic voltage regulator (AVR) [44].
- The voltage control of the secondary is an integrated control that is automatic to shorten the actions of local controllers. Hence, it is compact for the addition of reactive power inside a local power network.
- Tertiary voltage control refers to the standard optimization of reactive power flow to the power system.

Figure 8 outlines adequate voltage controls with one unit for generating a power system network. Voltage is controlled by the application and use of ratio-changing equipment such as transformer taps and reactive power control such as capacitors, reactors, static-VAR (Volt-ampere reactive) compensators, generators, and rarely synchronous condensers throughout the transmission system. The system operator controls and monitors the voltage and provides the grid's reactive power requirements. VAR absorption at low-load times is necessary to prevent the voltages from becoming too high. In contrast, VAR production can help prevent voltage levels from getting too low during high-load times [62]. In specific areas, it is more economical to purchase a reactive power support utility from a client or producer than to provide immediate reactive support. Dynamic or static VARs can be used to preserve or deliver the units (unit, primarily) based on its ability to keep the VAR output rapidly or faster retaining. In VAR, the cost of a dynamic device is much higher than a static device.

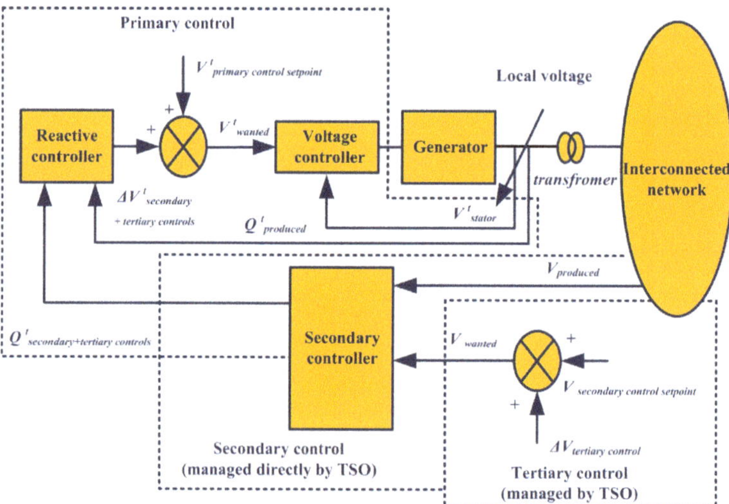

Figure 8. Three functional voltage controls for a generator.

Cost of Voltage Management

The cost of providing reactive power is mainly the cost of capital for equipment such as generators and capacitors, which can reduce the flow of capital as expected for backup. In addition, reactive support should be allocated to the operating costs of over or under the excitation of generating units. The main cost of generator voltage support is for rotor, stator, exciter, and step-up transformer losses. In some cases, the cost of the opportunity is associated with the loss of VAR production or absorption of real-time power-generation capacity. In the case of reactive power, the actual cost of producing a service can be mostly fixed costs and is usually lumped or isolated. In general, variable cost components are relatively low [43].

3.3.3. Capability of Black Start

The capability of black start has a characteristic of a black start station, which can start from a shutdown at any cost of its production units and integrate into the system to support a portion of the system and also be coordinated to the system under procedures without having an external power supply. Comparatively, it is termed as a method to energize a system from cases of total or fractional failure. Thus, it must have the generating units with black start ability [63]. The black start unit is measured as a generating unit, which can be started without power supply or determines the capacity with unit load to operate at a reduced level. Therefore, there is grid separation. It consists of at least one black start unit, such that a generating unit of black start reserve system would be able to energize the power system. Therefore, if there is a power outage in the system, it will provide power to local needs.

Electric supply is required to start all generating stations, except for a few small hydro stations. These stations have required some supply for work consumed in the auxiliary unit. So, the turbo generator is used to help in this process of operation. Usually, batteries provide this supply to small diesel generators. For large diesel generators, it is provided by pressurized air and diesel engines that deliver the start-up supply for gas turbines (start-up supply required for gas stations is usually 2% of installed capacity and 1% of hydro stations). Therefore, such a generating unit can be used to energize the subsystem/generating stations [57].

It is essential to satisfy the associated performance standards [63]:

- It can shut its circuit breaker for dead bus based on demand.
- It must be able to keep the frequency under various loads.
- It is capable of having a voltage supply for unstable loads.
- It is optimal to have an output rate within the given time as chosen by the system operator.

3.3.4. Inertia Response for RES

In general, when generation plants are connected to the network, RESs that are established as an alternative source of energy in the near future will pose new problems with the power supply quality. However, an adequately managed DG can be very useful in distributed networks to improve service continuity in specific geographical locations [64,65]. High penetration of RESs leads to critical challenges of frequency stability.

Firstly, RESs usually have inertial responses that are weak or non-existent. For instance, a power electronic converter typically connects a variable-speed wind turbine to a network that effectively dissipates wind turbine inertia from reducing transient systems. In addition, solar PV plants do not offer the power system with any inertia response. When traditional sources are replaced with RESs, the inertia of the entire electrical system is reduced [66]. Subsequently, an increase in the RES penetration rate reduces the number of manufacturing units that provide primary and secondary control reserve power. For this reason, as explained in [67] and shown in Figure 9, the frequency variance will be increased. The RESs need to create new frequency control techniques to enable them to participate in frequency regulation operations to overcome the frequency stability challenges posed by the small inertial response and reserve power.

Figure 9 shows the typical frequency response with operating limits. The machine frequency is about 50 Hz during normal operations. However, when an event triggers an imbalance in generation and demand, the frequency of the system begins to decline with the frequency levels, relying on the total system inertia and total unbalanced energy, as shown in the swing formula [68]:

$$\frac{df}{dt} = \frac{f_o}{2H_{sys}S_B}(P_m - P_e) \tag{1}$$

where df/dt is the frequency shift speed, H_{sys} is the maximum inertia constant of the unit, S_B is the rating of generator power, P_m and P_e, are the mechanical power and electrical power, and f_o is the frequency of the system. The synchronous generator releases kinetic energy stored in its spinning mass before any controller activation and up to ~10 s [69] due to the inertia reaction. After that, the primary frequency controller is activated immediately if the frequency deviation exceeds a certain value.

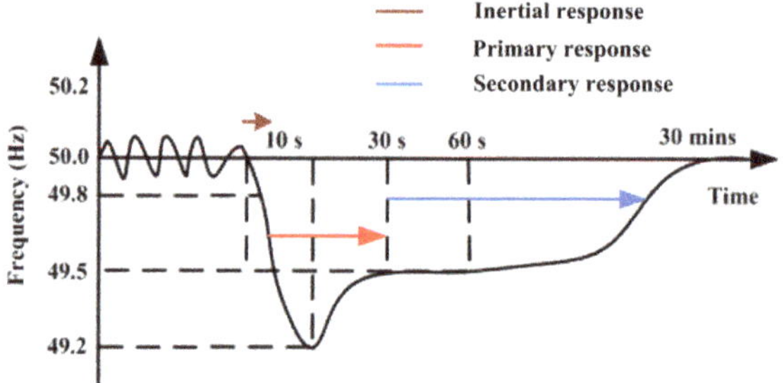

Figure 9. Time stages involved in responding to system frequency [67].

Using the generator governor, this controller returns the frequency to store values in 30 s [70,71]. To restore the system frequency to its nominal value, a new control called secondary control will be triggered after the 30 s. As illustrated in Figure 9, it takes several minutes for the secondary controller to restore the system frequency to its nominal value. Backup capacity is sufficient at this time to cover the increase in power demands. Finally, it enables the control of the tertiary frequency control by residual power deviation. Unlike primary and secondary controllers, the tertiary controller requires manual adjustment when generators are dispatched or scheduled timing adjustments [71]. Frequency and inertial control methods for RES are generally divided into two main categories: RES control techniques without any ESS assistance and RES control techniques with ESS. In Table 5, the frequency/inertia command merits and demerits for RES with and without ESS are illustrated.

Table 5. Merits and demerits of frequency/inertia control for RES (Renewable Energy Sources) with and without ESS [72].

ESS	Source Type	Methods	Merits	Demerits
Without	Solar	Deloading	Additional element is not required. Inertia and frequency regulation are provided.	It loses some energy percentage. It depends on the conditions of the environment.
	Wind	Inertial Response	Power obtained from the rotating mass directly.	The second drop in frequency may occur in losses.
		Deloading	Primary frequency control is provided.	It loses some energy percentage.
With	Solar	Deloading MPPT	The system is highly effective. Removes instabilities in power.	Higher cost due to the price of the battery and lose some energy. If the battery is fully charged, it fails to absorb power from the grid.
	Wind	Inertial response	The technique is highly reliable.	Compared to the above techniques, the value is quite higher. High battery price and energy costs.

3.3.5. FRT and Reactive Power Support

The radical climate variations caused by global warming and rising demand for energy have led to the development of renewable energy worldwide. Among other RES, wind energy is a pioneer and has grown both in terms of capacity and adopting technology. Wind power generation has changed over the past 30 years, including changes in electrical and mechanical systems, control methods used, and requirements for power system integration [73].

FRT capability during transient conditions and reactive power control during stable-state conditions pose considerable challenges for variable speed wind turbines [74] among the new grid code for wind energy grid integration. During faults and voltage sag conditions, wind farms are no longer allowed to disconnect but are expected to function as conventional power plants, provide the support for reactive power and be connected during system failures [75]. The grid code requirement (GCR) for FRT determines the time of fault (T_{fault}), permitted fault voltage (V_{fault}), recovery time (T_{rec}), voltage to be recovered within recovery time (V_{rec}), and the prescribed duration or settling time (T_{set}), as shown in Figure 10.

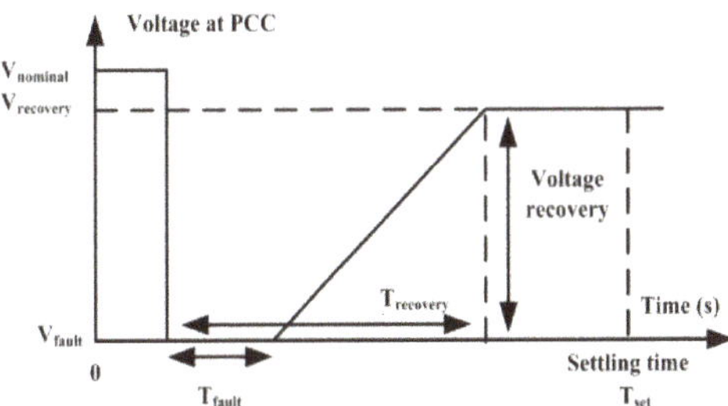

Figure 10. Basic FRT (fault ride-through) capability profile.

After clearance of the fault, T_{set} will indicate the voltage recovery to the nominal voltage (V_{nom}) prefault condition. The ability to return to the standard wind generator operating state assesses the determination of the power system under unstable conditions. The FRT requirements also require the increased injection of reactive current when the voltage is low, as shown in Figure 11. In response to severe voltage drop, reactive current assistance should be achieved within 20 ms after failure is detected.

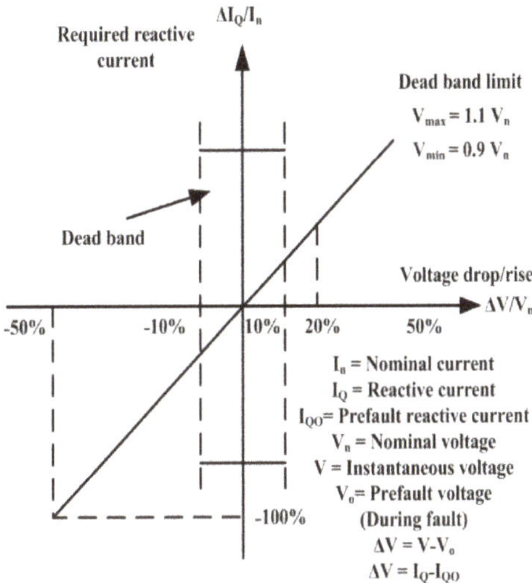

Figure 11. Required reactive current profile based on grid code of E.ON [76].

Double-fed induction generator wind turbines (DFIGWTs) are popular for WT technology. DFIG is an innovative type of variable wind speed generator. However, DFIGWTs can cause disturbances in the grid voltage. They use both the rotor-side converter (RSC) and the control of the grid-side converter (GSC) [77]. The RSC controls maximum power capture by control of rotor rpm and GSC controls the active and reactive power supplied and maintains a constant DC-link voltage. The voltage at the fault event also fails to zero and the active power output decreases, which rapidly increases the current of the rotor in an attempt to compensate for the active power of the RSC. The converter then raises the voltage of the rotor, resulting in a high voltage in the DC connection, and DFIG rapidly loses internal magnetization in proportion to the voltage loss. Demagnetization induces high out-rush/over-current currents on both stator and rotor circuits that are greater than the converter's ratings [78]. This leads to the WT connected to the grid being tripped.

Consequently, the GCR suggests the ability of the FRT to ensure continuous operation and prevent excessive power loss due to defects, promoting grid recovery during fault and minimizing problems with resynchronization after fault clearing by reactive power support [79]. The merits and demerits of protection methods during the FRT process are presented in Table 6.

Table 6. Merits and demerits of control techniques using for FRT.

Reference	Using Technique	Merits	Demerits
[80]	Crowbar	Activated in the event of failures and prevents RSC from overload.	When crowbar is applied, RSC control is lost.
[81]	SGSC	Damping synchronous stator frame flux oscillations and allowing the stator flux variable to be handled directly.	Weaknesses in preserving the power balance of the DC-link.
[82]	ESS	Improves DFIG's transient dynamics and power systems' transient stability. DFIG's steady-state active power output is regulated.	Battery unit operation and maintenance issues. Loss of stored energy in the form of self-discharges when not in use.
[76]	MSDBR	This method prevents the use of both the crowbar and the DC chopper. Series compensation system and includes power evacuation.	The injection efficiency of reactive energy is not yet studied. Compared to the above techniques, the value is quite higher.

3.3.6. ESS in Congestion Management and Economical Scheduling

Existing congestion management systems (CMSs) typically determine the transmission point day-ahead stage, the generation process, and the ability to anticipate demand at this level [83]. High penetration of DER in smart distribution systems may lead to network congestion and voltage problems. The establishment of the power distribution market is a good tool for effectively and efficiently managing large amounts of DERs in distribution networks. Active customers and utilities can use this distribution-level market to realize their potential value in distribution grids. A well-designed market could also encourage rational deployment and investment in future planning of new DERs. Therefore, the main goal of market integration is to develop international markets in real time, accepting all network limitations [84]. Therefore, effective CMS should be fully integrated with intraday and stabilization market design. As mentioned earlier, the following are the five key elements to establishing a dynamic CMS:

- Efficient national CMS and integrated with international CMS to get complete utilization of current transmission capacity.
- Combined distribution of global transmission capability, for the flexible usage of transmission capacity where it is more required at a day-ahead level.
- The day-ahead energy market integrated with transmission allocation is the ability to make complete usage of low-cost distribution possibilities.
- The flexible operation across the power system, improvement in RES forecasts such as solar/wind, and other uncertainties in a day is possible with the integration of transmission distribution with the day-ahead energy market.

A precise method for CMS accepts the efficient support and reliable evaluation of future congestion designs for private and public outcome makers to monitor the investment options. The day-ahead economic scheduling of ESS is the regulating of electricity market positions. The ESS integrates with residential customers in a low voltage generation grid and PV generation, whereas the objective is to expand the revenues from energy arbitrage in day-ahead auction [85]. The internal model of the ESS design addresses the charging/discharge process limitations. Therefore, the PV generation, residential load, price of the market, and formulated constrained optimization challenge minimize the cost function. It mainly focuses on system efficiency, which significantly affects economic performance due to energy losses in the period of charging/discharging cycle.

The integrated PV and ESS play an important role in demand-side management (DSM) activities, and these improvements can influence the performance of electric power systems (EPSs). The market can exist in the wholesale trade of electricity segment and it was developed for energy exchange to handle the day-ahead auction markets. Thus, ESS combining with uncertainty sources for day-ahead economic optimization is a compelling research topic in the economic scheduling operation process. In [86], the authors proposed an hourly discretized optimization algorithm for a residential consumer, PV generation, and a distributed BESS. In this context, the aim was not only on the residential generation, but also on regional stages, where the BESS performed for day-ahead scheduling can improve revenues from energy arbitrage and is subject to the technical constraints of the supply network and storage system. The summary of the MG components and corresponding ancillary services are illustrated in Table 7 [87].

Table 7. Summary of corresponding ancillary services for MG components.

MG Components	Ancillary Services To Main Grid
All DERs, WTGs, PV systems, hydro power plants, and loads with ESSs units but not thermal-driven CHP	Frequency regulation
Inverter and SG coupled DG/ESS units and loads but not IG coupled DG	Voltage control, CMS, Optimization of grid losses
WT's coupled with inverters, SGs, PV with inverter, Micro-hydro with inverter/SG and ESS	Black start
WT's with DFIG/Inverter, PV with Inverter, Micro-hydro with Inverter, CHP with inverter, ESS	FRT Capability

Further approaches based on the location of the economic evaluation of the specified application and the feasibility study of the ESS were investigated in [88]. The authors illustrated in [89] complete analysis for a day-ahead market of the European power exchange group, an exchange for power spot transactions in Germany, Switzerland, Austria, and France. This study evaluated the corresponding economic evaluation for the possibility of energy arbitrage function in global markets. The key drivers for improvement of ESS solutions are the market prospects over the provision of ancillary services; energy arbitrage; improving the efficiency of transmission, generation, and distribution; and several balancing applications.

3.3.7. Energy Management System

To assess an ancillary service technology of a MG, first the grid connection technology and then the entire controlled DER must be analyzed [87]. The economic framework for supplying ancillary services through MG is based on the energy management system (EMS) decision-taking capability, which provides the most promising technology for ancillary services, the most feasible service for community MG owners to generate incremental revenues, and provides security of supply to customers.

As shown in Figure 12, this EMS controls MG power flows through adjustment of power imports/exports from/to the main grid, regulates dispatchable DERs and controls loads based on current information, the generation, and loads to achieve certain operational targets (for example cost reduction) and optimizations of the end of the market. In the case of many ancillary services, the EMS determines whether and at what price it will provide the service in a day-to-day market. The EMS then bids into the market and sees whether the bid is successful. If this service is successful, EMS plans to provide it the next day. Therefore, MG owners can benefit from the ancillary services market as the main source of income. MG owners benefit from the energy markets in many parts of the world. Finally, Table 8 summarizes the current and possible ESS facilities for different application areas.

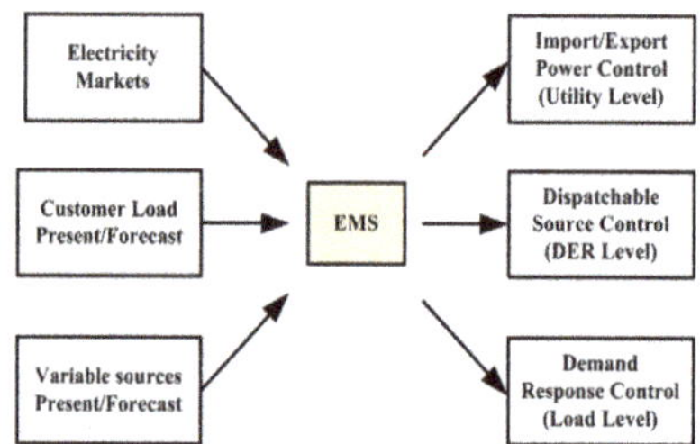

Figure 12. Functions of EMS (energy management system) for MG.

Table 8. Overview of current and possible ESS options with their specifications for different applications [90].

Application Area	Summary	Characteristics and Specifications	ESS Technology Options
Power quality	The issue of Power Quality (PQ) is one of MG's major technical challenges. The PQ level of the MG network must be analyzed and quantified to provide a better PQ of the energy provision. In both the on grid and off grid mode of MG operation, voltage and frequency variation are analyzed under different generation and load conditions. In order to achieve a better quality of power supply in the MG system, the level of PQ impact in the MG network must be quantified in various scenarios.	~<1 MW, Response time: ~ms, Discharge period: ms to s	Exp: FES, BES, SMES SCs; Pro: FBs
RES power integration	The intermittent generation of renewables can be backed up, stabilized, or supported by integration with ESS.	~100 kW–40 MW < 1 MW, Response time: ~s to min, Discharge period: up to days	Exp: FES, BES; Pro: PHES, CAES, FCs
Frequency control	Based on active power control by controlling the DER output. Generation is adjusted to load minute by minute to maintain a specific system frequency in the control area. The micro-sources (DGs) of MG connected to the grid and located close to the load pockets are an effective way of delivering this service.	Up to MW level Response time: ~s, Discharge period: s to min	Exp: BES, FBs, CAES Pro: FES, SCs
Voltage control	EPSs dynamically respond to changes in active and reactive power, thereby influencing the voltage profile and magnitude of the networks. Dynamic voltage behavior control can be improved with the functions of ESS facilities. Various ESS technologies can be used effectively for voltage control solutions.	Up to few MW level, Response time: mins Discharge period: Up to mins	Exp: BES, FBs; Pro: SMES, FES, SCs
Spinning reserve	ESSs have spinning reserve functions if the generation (or load decrease) increases rapidly enough to lead to contingency. ESS units should be able to react immediately and to keep outputs up to a few hours.	Up to MW level, Response time: s Discharge period: 30 min to few hrs	Exp: BES, Pro: FCs, FBs, FES, CAES, SMES
Load levelling	Load-levelling is a way to balance large fluctuations in electricity demand. Traditional batteries and FBs should reduce overall costs and improve cycling time with peak shaving applications as well as in load following and time-shifting.	Several hundreds of MW level, Response time: mins Discharge period: ~12 h and even more	Exp: BES, PHES, CAES; Pro: FCs, FBs, TES
FRT capability	There has been much interest in the concept of MGs recently. As the power capacity of MGs increase, EPS can deliver significant power from DGs. During power grid interruptions, a high-powered MG disconnect can lead to power grid instability. New grid codes that address stringent requirements. However, broadly linking MGs through distribution networks requires a change in their philosophy of connecting them to the utility grid. Grid-connected MG requires FRT capabilities and ancillary services during abnormal grid operations.	~100 kW–100 MW Response time: Up to ~s, Discharge period: s to mins and even hrs	Exp: BES, FBs, CAES; Pro: FCs, FES, SCs

Table 8. *Cont.*

Application Area	Summary	Characteristics and Specifications	ESS Technology Options
Transmission and distribution stabilization	To control power quality, reduce congestion, and/or ensure that the system operates under normal working conditions, ESS can be used to synchronize the operation of a power transmission line or parts of a distribution unit. Such applications require immediate response and a relatively large grid demand power capacity.	Up to 100 MW level, Response time: ~ms Discharge period: ms to s	Exp: BES, SMES; Pro: FBs, FES, SCs
Black-start	ESS can deliver a system from a shutdown condition to its start-up without using electricity from the grid.	Up to ~40 MW level, Response time: ~mins, Discharge period: s to h	Exp: BES, CAES, FBs; Pro: FCs, TES
Standing reserve	ESS facilities serve as temporary additional generating units in the middle to large scale grid to balance power supply and demand at a certain time. The standing reserve can be used to meet current demand that is higher than future demand and/or plant failure.	1–100 MW level, Response time: <10 min, Discharge period: ~1–5 h	Exp: BES; Pro: FCs, FBs, PHES, CAES
Load following	ESS installations can support subsequent electricity demand load changes. The Irvine Smart Grid Demonstration test project with advanced batteries offers load follow-up and voltage support services in California.	Up to hundreds of MW level, Response time: up to ~1 s, Discharge period: min to few hours	Exp: FBs, BES, SMES; Pro: FCs
EMS	In EMS, ESS plays an important role in optimizing the use of energy, and decoupling generation time and energy consumption. Typical EMS applications are time-shifting and peak shaving.	>100 MW for large scale, ~1–100 MW for medium/small scale Response time: mins, Discharge period: hrs to days	Exp: Large—HS, CAES, TES; Smal—BES, FBs, TES Pro: FCs, FES
Time-shifting	It can be attained by stored electrical energy when it is cheaper, and the stored energy used or sold during periods of high demand.	~1–100 MW & even more Response time: mins, Discharge period: ~3–12 h	Exp: PHS, CAES, BES; Pro: FBs, FCs, TES
Peak shaving	Peak shaving is the use of stored energy during off-peak periods to offset energy generation over maximum power demand periods. The ESS function offers economic benefits by reducing the need to use high-cost electricity generation.	~100 kW–100 MW & even more Response time: mins, Discharge period: hr level, ~<10 h	Exp: PHS, CAES, BES Pro: FCs, TES
Network stability	Some grid/network power electronic, information and communication systems are highly vulnerable to fluctuations in power. ESS installations can provide the protective function for these systems, requiring high ramp power and high cycling time capabilities with a rapid response time.	Up to MW level, Response time: ms, Discharge period: Up to ms	Exp: BES FES, SCs, SMES; Pro: FBs

s: seconds, ms: milliseconds, min: minutes, h: hours, Exp: Experienced, Pro: Promising.

The evolution and development of ESS technology will be possible with advanced control methods to mitigate the abovementioned problems. It not only depends on the enhancement in characteristics of energy storage, system control, and management strategy, but also requires the cost minimization and supports for long-term supplies, a positive stable market, and a plan to control and sustain the healthy development of the ESS industry.

3.4. Global Prospects on Ancillary Services

Ancillary services usually include operational reserve, regulation of frequency, and also considerably faster response. Great Britain operators include the mandatory ancillary services such as frequency response, black start, reactive power, load following, reserve capacity, and demand turn up and intertrip [91]. The ancillary services are mostly supported by power plants or large, pumped, hydro storage systems [92]. It was found that there is no standardized classification of ancillary services. Several ancillary services are recognized around the globe based on the power system network and regional requirements. As per Eurelectric [93], the frequency and voltage control and system stability can be considered as ancillary services. It has been reported by CIGRE (International Council on Large Electric Systems) that voltage, frequency control, network control, and black start are the required ancillary services [94]. FERC (Federal Energy Regulatory Commission) in the U.S. describes the need for services to support the production of electricity from seller to consumer, while maintaining a consistent process of coordinated communication. The FERC recognizes six vital ancillary services that should be included in open access transmission tariff [95]: (1) scheduling, system control, and dispatch, (2) frequency regulation, (3) energy imbalance, (4) volt/VAR (Volt-ampere reactive) control from generation sources, (5) operating reserve-spinning reserve, and (6) operating reserve-supplemental reserve.

The Canadian electricity board specified that the services are mandatory to control for operation of the interconnected generation system and that the voltage and frequency regulations require acceptable control limits [96]. Table 9 summarizes some of the popular ancillary services available in the main and large power systems of the world. Power systems in many countries are operated by similar or different operators at the regional level. Therefore, there may be different set/subset services in different parts of the same country.

Table 9. Available ancillary services around the world.

Ref.	AS/Country	SR	TC	TMR	VS	DTI	BS	SuR	OR	SR	EB	FC	S/D	VARC	VC
[97]	CN						✓			✓		✓	✓	✓	✓
[71]	US							✓		✓	✓	✓	✓	✓	✓
[98]	IN						✓		✓	✓		✓	✓	✓	✓
[42]	RU						✓				✓	✓		✓	✓
[99]	JP				✓		✓		✓	✓	✓	✓	✓	✓	✓
[71]	CD			✓		✓	✓			✓		✓		✓	✓
[100]	GE	✓			✓		✓	✓			✓	✓	✓		✓
[101]	BR							✓	✓	✓		✓			✓
[102]	SK						✓	✓		✓		✓	✓	✓	✓
[103]	FR	✓					✓		✓			✓		✓	✓
[104]	AU	✓					✓			✓	✓	✓	✓	✓	✓
[105]	UK					✓	✓								✓
[106]	MX	✓	✓	✓	✓	✓	✓	✓	✓	✓	✓	✓	✓	✓	✓
[107]	IR						✓		✓			✓	✓	✓	✓
[108]	SA	✓					✓			✓		✓	✓	✓	✓
[109]	TR						✓	✓	✓		✓	✓	✓	✓	✓

Ref., reference; AS, ancillary services; SR, system restoration; TC time correction; TMR, transmission must run; VS, voltage stability; DTI, demand turn-up and intertrip; BS, black start; SuR, supplemental reserve; OR, operating reserve; SR, spinning reserve; EB, energy balancing; FC, frequency control; S/D, scheduling or dispatch; VARC, VAR control; VC, voltage control.

4. Drivers Involved in MG Development and Deployment

Earlier, MG was described by the Consortium of Electric Reliability Technology Solutions (CERTS) as a semi-autonomous collection of DGs and controllable loads simultaneously producing a secure and reliable operation to the local community network [110]. MGs incorporate modular DERs such as wind, solar PV, and fuel cells to form a low-voltage distribution system with the help of storage devices and controllable loads [111].

4.1. Functions of MG

The specific definition of MG cannot be determined easily because the scope and structure of MG may vary. MG is a limited group of multiple low- or medium-voltage distributed generation, load, and storage units that act as a self-coordinated system. MGs have the role of self-healing smart grids in the future as they have an islanding capability to reduce outages. They consist of interrelated, renewable, and conventional energy sources and are often associated with the distribution grid at a point of common coupling (PCC). Therefore, for the DSO (Distribution System Operator) perspective, they can occur as a single, flexible, and controllable entity. There is no definition of the size of MGs, and this may vary, but it is usually considered to be a small part of low or medium voltage distribution network [112]. The process of market participation for MG is shown in Figure 13.

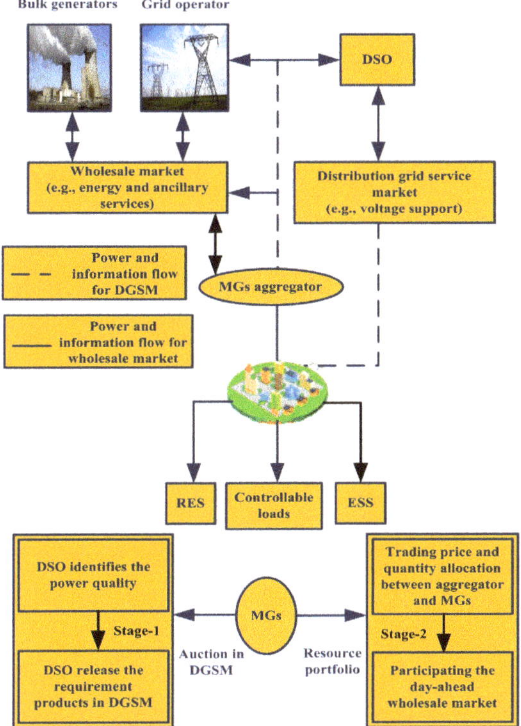

Figure 13. Market participation process for MGs [113].

MGs are key components of the local energy markets and offer the possibility of peer-to-peer energy transactions [114]. To operate, MG requires a variety of components. In simple terms, the foundation of MGs is DERs. This includes DG, storages, or active loads. Second, a physical network is required to connect all DER and consumers. To manage and control distributed energy flows delivered through integrated communication, advanced control and demand response technology is required. The ownerships of MGs can be divided into three different designs. In "DSO monopoly MG", the DSO owns the grid and all the responsibilities and costs of operating the MG come to the operator, but also the benefits. Usually, this occurs in nonliberalized markets, where the DSO includes energy distribution and retail. "Prosumer Consortium MG" has single or multiple customers with different DERs, thus benefiting from lower electricity bills or revenues generated when supplying the grid with surplus electricity. "Free Market MG" is run by different stakeholders such as DSOs, prosumers, or consumers. It means that a central controller has functioned MG. Subsequently, this type of MG is driven by many stakeholders, as well as the profits shared among them [115].

4.2. Factors Responsible for MG Development

There are three distinct benefit classes related to MGs: technical, economical, and environmental. From a technical view, some of the benefits include the power distribution to remote communities such as energy enhancement, providing communication support in remote areas, and it will help to reduce the huge power blackouts for networks [40,41]. Commercial interests have been extensively studied [42]. They describe emissions, line losses, customer disruption costs, and fuel price reductions. Conversational support of MGs has been addressed in [43], with MGs ending with less discharge of pollutants and greenhouse gases. Also, the generation system needs a smaller physical footprint. MG consumption also increases the number of clean energy sources integrated into the grid. This reduces the dependence on external sources of energy. The aspects of MG driving fall into three broad categories as described in Table 10, below:

- Energy safety measures,
- Economic gains, and
- Clean energy integration.

Table 10. Classification of MG drivers.

Reference	Category	Driver	Outline	Current Examples
[29]	Energy Security	Severe weather	It is a known fact that weather might be a greater disruption, especially in countries like the United States. This is the reason that climate change will result in a need to address the resilience of the grids. Thus, MGs could offer power to major services and groups through their spread generation assets if the main drop.	Costs levied on-grid outage concerning weather-related issues in the U.S. alone between 2003–12 ranging around $18B–$33B in a year due to poor output and wages disposal, also from spoiling inventory, delayed production followed by losses from the electric grid [29].
[32,116]		Outages	Electrical grids of critical capacity remained a mild issue in a system that can result in a domino effect that takes down a complete electrical grid [32]. MGs reduce this risk by dividing the grid into minor functional units, which can be isolated and operated independently whenever needed.	The U.S. Northeast Blackout of August 2003 made nearly 50 million people suffer because of 61,800 MW of load reduction [116].
[117–125]		Physical and Cyber outbreaks	Today, the grid depends on progressive information and communications technologies, thus making it susceptible to cyber-attack [117]. The central grid network involves larger components, which are rather costly and difficult to exchange whenever they get damaged. MGs, with the decentralized design, are less susceptible to outbreaks on distinct sections of generation or transmission power supplies, natural [118,119], artificial, or electromagnetic pulse incidents might also under disastrous results [120,121].	Ukrainian cyber-attacks [122] in 2015 and Israel in 2016 were effectively eliminated [123]. Larger transformers were confronted at a major California substation in 2013 [124,125].
[126–128]		Saving the cost of infrastructural facilities	U.S. electricity grid systems were not able to keep up with the generation pace. Consequently, the capacity of the grid is inhibited in several zones, and components are relatively old, with 70% transmission lines and transformers now moving forward to 25 years. The age of the power plant is over 30 years old [126]. It has the capability of avoiding or deferring investments for replacement.	The deferred construction over $1B substation from Queens and the Brooklyn area of NY [127]. Costs levied $40,000–$100,000 per mile, relying based on prominent factors like terrain, design, and cost of labor of building new primary distribution systems [128].
[129,130]		Fuel Savings	MGs provide various efficiency types, including minimizing losses in the line, the combination of heat, cooling, and power losses, along with the shift to distribution systems of direct current to remove unnecessary DC-AC conversions. When absorption cooling technology with the combination of heat and power applications might aid in addressing the peak electricity demand that usually occurs in the summer season [130].	The losses from wastage in transmission and distribution are about 5% & 10% over a gross electricity generation [129]. When appropriately used, the effectiveness of heat and power systems can reach 80–90% [108], which is found to be much higher than the average efficiency of the U.S. grid that is currently (only ~30–40%) used [129,130].
[131–135]		Ancillary Services	Conventional ancillary services consist of relief from congestions, regulation of frequency & load, black start, controlling both reactive power & voltage along with spinning supplies. This is because of their capability to provide the same inertia as that of a conventional power generation system, non-spinning, and additional reserves [131,132]. Also, all the individual operations should be included in the list [133].	Current rulings under 755 & 784 of U.S. FERC necessitate the fast-reacting reserves that are employed in MGs that needs to be compensated as per their speediness and accurateness, options for the possibility of new revenue system [134,135].
[136–141]	Integration of the clean energy system	Need to secure inconstant and uncontrollable resources	Significant sources for clean energy sources for addressing climate change such as solar PV and wind are variable and non-controllable that could result in challenges such as excessive generation [136], steep ramping [137,138] and voltage control [139,140] MGs are designed for handling variable generation by making use of storage technologies for locally balancing the generation of loads.	In Texas, California, and Germany, the cost of electricity is relatively high, which reflects the imbalance found between demand and supply [140,141].

4.3. Application of MG

To integrate DERs, the smart MG is the most effective platform, for example, solar PV, diesel generator, fuel cell, wind turbine, and micro combined heat and power sections. Undefined problems arise from RES for sustainability development at different nodes of MG grid delivery deployed using storage. An ESS utility includes a power-quality controller, which produces the active and reactive power required by clients. The principle of controlling voltage for high loads with load shedding is not sufficient to achieve a higher quality of power supply, which has an effect on MG [99].

Therefore, this is an additional advantage of ESSs, which often have the effect of improved energy quality in the MG environment when a fault frequently occurs [142,143]. In addition, ESSs offer a variety of applications [144] such as black start [145,146], power alternation inhibiting [147,148], grid inertia response [149], wind power gradient reduction [150], peak shaving [151,152] and load following [153]. Nowadays, researchers are trying to come up with a number of approaches to develop power management and system stabilization of MGs by using ESS. To integrate resources for MG, storage systems and the power electronic interfaces are discussed in [154]. A cooperative control mechanism for the control of frequency and voltage is discussed in [155], which also helps to reduce the power variation from RESs (such as wind turbines) and stabilize the frequency in the control system and in a wide range of power systems. This suggests the effect of the electromechanical oscillations of the fast response of energy storage in the power system [156]. This helps reduce the short-term variations and minimizes the capacity of the storage system.

5. Conclusions

MG technologies have been serving as a key spot in research over distributed energy systems. The core of the continuing development of these services is to ensure that effective compliance systems are in place for security measures, scheduling followed by the design of MG, as well as related issues with regulating and EMS operation for control strategies and ancillary services. The monitoring and evaluation system helps to plan the potential direction for the implementation of new ancillary services as competition grows in this sector. In addition, the underlying prospects of governing ancillary services for smart MG remain a topic of research in detail in this paper. There is a rapid reduction in battery storage costs and solar PV production for a moment close to cost equality with conventional power sources. Therefore, the larger implementation of these systems has accelerated for the assumption of energy at some point, where other customs are concerned with the import and export of electricity with the end user. Previously, DERs were related to the electrical grid, and it was sufficient for the community to identify and design which architectural design would best integrate with distributed technologies. Instead of ensuring the reliability, protection, and stability of renewable sources in microgrids, installing ESSs can also generate substantial revenues through ancillary services to the main grid. With the use of ESS for ancillary services, the overall deployment cost of linked microgrids is decreased. ESS plays an important role in every sector, ensuring the safe, stable, reliable operation of power systems and having a wide range of application capabilities. Smart MGs will achieve this transformation through demand and supply balance, while ensuring reliability and flexibility, as opposed to increasing natural and human-made obstacles. Through an increasing number of small-scale MGs, and by providing the system with the state-of-the-art multimarket framework, it encompasses a large part of the planning process for the real environment. Due to quick response and stability control in MGs, such as voltage and frequency regulation, reactive power injection, supply balance, and demand response, the results of MG can be developed in the market interest of future research directions.

Author Contributions: Conceptualization, methodology, investigation, resources, data curation, writing—original draft preparation, G V B.K.; visualization, supervision, review, and editing, K P. Both authors have read and agreed to the published version of the manuscript.

Funding: This work is supported by the Department of Science and Technology (DST), Government of India (GOI) with the project grant SR/FST/ETI-420/2016(C) under FIST scheme.

Conflicts of Interest: The authors declare no conflict of interest.

Abbreviations

MG	Microgrid
PV	Photovoltaic
RES	Renewable Energy Sources
DER	Distributed Energy Resources

ESS	Energy Storage System
DGDG	Distributed Generation
DSO	Distribution System Operator
SGSC	Series Grid Side Converter
VAR	Volt-ampere reactive
FERC	Federal Energy Regulatory Commission
NERC	North American Electric Reliability Corporation
CIGRE	International Council on Large Electric Systems
UCTE	Union for the Coordination of Transmission of Electricity
MSDBR	Modulated Series Dynamic Breaking Resistor
CN	China
IN	India
JP	Japan
GE	Germany
SK	South Korea
US	United States
RU	Russia
FR	France
UK	United Kingdom
BR	Brazil
IR	Iran
SA	South Africa
MX	Mexico

References

1. Levron, Y.; Guerrero, J.M.; Beck, Y. Optimal Power Flow in Microgrids With Energy Storage. *IEEE Trans. Power Syst.* **2013**, *28*, 3226–3234. [CrossRef]
2. Liu, X.; Wang, P.; Loh, P.C. A Hybrid AC/DC Microgrid and Its Coordination control. *IEEE Trans. Smart Grid.* **2011**, *2*, 278–286. [CrossRef]
3. Duan, C.C.S.; Liu, T.C.B. Smart energy management system for optimal microgrid economic operation. *IET Renew. Power Gen.* **2011**, *5*, 258–267. [CrossRef]
4. Guerrero, J.M.; Chandorkar, M.; Lee, T.; Loh, P.C. Advanced Control Architectures for Intelligent Microgrids — Part I: Decentralized and Hierarchical Control. *IEEE Trans. Ind. Electron.* **2013**, *60*, 1254–1262. [CrossRef]
5. Molina, M.G. Distributed Energy Storage Systems for Applications in Future Smart Grids. In Proceedings of the 2012 Sixth IEEE/PES Transmission and Distribution: Latin America Conference and Exposition (T&D-LA), Montevideo, Uruguay, 3–5 September 2012; pp. 1–7. [CrossRef]
6. Force, I.T.; Olivares, C.D.E.; Mehrizi-sani, A.; Etemadi, A.H.; Cañizares, C.A.; Iravani, R.; Kazerani, M.; Hajimiragha, A.H.; Gomis-bellmunt, O.; Saeedifard, M.; et al. Trends in Microgrid Control. *IEEE Trans. Smart Grid.* **2014**, *5*, 1905–1919. [CrossRef]
7. Martin-martínez, F.; Rivier, M.A. literature review of Microgrids: A functional layer based classification. *Renew. Sustain. Energy Rev.* **2016**, *62*, 1133–1153. [CrossRef]
8. Gundumalla, V.B.K.; Eswararao, S. Ramp Rate Control Strategy for an Islanded DC Microgrid with Hybrid Energy Storage System. In Proceedings of the 2018 4th International Conference on Electrical Energy Systems (ICEES), Chennai, India, 7–9 February 2018; pp. 82–87. [CrossRef]
9. Hirsch, A.; Parag, Y.; Guerrero, J.M. Microgrids: A review of technologies, key drivers, and outstanding issues. *Renew. Sustain. Energy Rev.* **2018**, *90*, 402–411. [CrossRef]
10. DeBlasio, D. Toward a self-healing smart grid. *Fortnightly Mag.* 2013. Available online: https://www.fortnightly.com/fortnightly/2013/08/toward-self-healing-smart-grid (accessed on 20 September 2016).
11. Asmus, P. Microgrids, virtual power plants and our distributed energy future. *Electr. J.* **2010**, *23*, 72–82. [CrossRef]
12. Madureira, A.G.; Peças Lopes, J.A. Ancillary services market framework for voltage control in distribution networks with microgrids. *Electr. Power Syst. Res.* **2012**, *86*, 1–7. [CrossRef]

13. *Distributed Energy Resources Roadmap for New York's Wholesale Electricity Markets*; New York Independent System Operator: New York, NY, USA, 2017.

14. Huang, J.; Jiang, C.; Rong, X. A review on distributed energy resources and Micro Grid. *Renew. Sustain. Energy Rev.* **2008**, *12*, 2465–2476.

15. Nick, M.; Cherkaoui, R.; Paolone, M. Optimal Allocation of Dispersed Energy Storage Systems in Active Distribution Networks for Energy Balance and Grid Support. *IEEE Trans. Power Syst.* **2014**, *29*, 2300–2310. [CrossRef]

16. Chen, Y.; Hu, M. Balancing collective and individual interests in transactive energy management of interconnected micro-grid clusters. *Energy* **2016**, *109*, 1075–1085. [CrossRef]

17. Muruganantham, B.; Gnanadass, R.; Padhy, N.P. Challenges with renewable energy sources and storage in practical distribution systems. *Renew. Sustain. Energy Rev.* **2017**, *73*, 125–134. [CrossRef]

18. Coelho, V.N.; Cohen, M.W.; Coelho, I.M.; Liu, N.; Guimarães, F.G. Multi-agent systems applied for energy systems integration: State-of-the-art applications and trends in microgrids. *Appl. Energy* **2017**, *187*, 820–832. [CrossRef]

19. Walton, R. Former FERC Chair Says Microgrids Are Key to Grid Security, Util Dive. 2014. Available online: http://www.utilitydive.com/news/former-ferc-chair-says-microgrids-are-key-togrid-security/327814 (accessed on 13 September 2016).

20. Center for Energy, Marine Transportation and Public Policy at Columbia University. *Microgrids: An Assessment of the Value, Opportunities and Barriers to Deployment in New York State*; New York State Energy Research and Development Authority: Albany, NY, USA, 2010.

21. Tweed, K. New York Looks to Cement Its Lead as Microgrid Capital of the World. 2015. Available online: https://www.greentechmedia.com/articles/read/new-york-looks-to-cement-itslead-as-microgrid-capital-of-the-world (accessed on 3 March 2016).

22. Pachauri, R.K.; Allen, M.R.; Barros, V.R.; Broome, J.; Cramer, W.; Christ, R.; Church, J.A.; Clarke, L.; Dahe, Q.; Dasgupta, P.; et al. Climate Change: 2014: Synthesis Report. Contribution of Working Groups I, II and III to the Fifth Assessment Report of the Intergovernmental Panel on Climate Change. IPCC. 2014. Available online: https://epic.awi.de/id/eprint/37530/ (accessed on 1 November 2014).

23. Gamarra, C.; Guerrero, J.M. Computational optimization techniques applied to microgrids planning: A review. *Renew Sustain Energy Rev.* **2015**, *48*, 413–424. [CrossRef]

24. Mariam, L.; Basu, M.; Conlon, M.F. Microgrid: Architecture, policy and future trends. *Renew. Sustain. Energy Rev.* **2016**, *64*, 477–489. [CrossRef]

25. Díaz-gonzález, F.; Sumper, A.; Gomis-bellmunt, O.; Villafáfila-robles, R. A review of energy storage technologies for wind power applications. *Renew. Sustain. Energy Rev.* **2012**, *16*, 2154–2171. [CrossRef]

26. Dragicevic, T.; Vasquez, J.C.; Guerrero, J.M.; Skrlec, D. Advanced LVDC Electrical Power Architectures and Microgrids: A step toward a New Generation of Power Distribution Networks. *IEEE Electrif. Mag.* **2014**, *2*, 54–65. [CrossRef]

27. Akorede, M.F.; Hizam, H.; Pouresmaeil, E. Distributed energy resources and benefits to the environment. *Renew. Sustain. Energy Rev.* **2010**, *14*, 724–734. [CrossRef]

28. Mekhilef, S.; Saidur, R.; Safari, A. Comparative study of different fuel cell technologies. *Renew. Sustain. Energy Rev.* **2012**, *16*, 981–989. [CrossRef]

29. U.S. Department of Energy's Offic. *Economic Benefits of Increasing Electric Grid Resilience to Weather Outages, Executive Office of the President*; 2013. Available online: https://www.energy.gov/sites/prod/files/2013/08/f2/Grid%20Resiliency%20Report_FINAL.pdf (accessed on 15 December 2020).

30. Hossain, E.; Kabalci, E.; Bayindir, R.; Perez, R. A Comprehensive Study on Microgrid Technology. *Int. J. Renew. Energy Res.* **2014**, *4*, 1094–1107.

31. May, G.J.; Davidson, A.; Monahov, B. Lead batteries for utility energy storage: A review. *J. Energy Storage* **2018**, *15*, 145–157. [CrossRef]

32. Brahmendra Kumar, G.V.; Palanisamy, K. A Review on Microgrids with Distributed Energy Resources. In Proceedings of the 2019 Innovations in Power and Advanced Computing Technologies (i-PACT), Vellore, India, 22–23 March 2019; pp. 1–6. [CrossRef]

33. Suvire, G.O.; Mercado, P.E.; Ontiveros, L.J. Comparative Analysis of Energy Storage Technologies to Compensate Wind Power Short-Term Fluctuations. In Proceedings of the 2010 IEEE/PES Transmission and Distribution Conference and Exposition: Latin America (T&D-LA), Sao Paulo, Brazil, 8–10 November 2010; pp. 522–528. [CrossRef]
34. Rosen, M.A.; Koohi-Fayegh, S. A review of energy storage types, applications and recent developments. *J. Energy Storage* **2020**, *27*, 101047. [CrossRef]
35. Kirubakaran, A.; Jain, S.; Nema, R.K. A review on fuel cell technologies and power electronic interface. *Renew. Sustain. Energy Rev.* **2019**, *13*, 2430–2440. [CrossRef]
36. Saraiva, J.T.; Gomes, M.H. Provision of Some Ancillary Services by Microgrid Agents. In Proceedings of the 7th International Conference on the European Energy Market, Madrid, Spain, 23–25 June 2010; pp. 1–8. [CrossRef]
37. Lasseter, R.H. MicroGrids. In Proceedings of the IEEE Power Eng. Society Winter Meeting, New York, NY, USA, 27–31 January 2002; pp. 305–308. [CrossRef]
38. Venayagamoorthy, G.K.; Sharma, R.K.; Gautam, P.K.; Ahmadi, A. Dynamic energy management system for a Smart Microgrid. *IEEE Trans. Neural Networks Learn. Syst.* **2016**, *27*, 1643–1656. [CrossRef] [PubMed]
39. Su, S.; Li, Y.; Duan, X. Self-organized criticality of power system faults and its application in adaptation to extreme climate. *Chin. Sci. Bull.* **2009**, *54*, 1251–1259. [CrossRef]
40. Kumar, A.; Chowdhury, S.P.; Chowdhury, S.; Paul, S. Microgrids: Energy management by strategic deployment of DERs—A comprehensive survey. *Renew. Sustain. Energy Rev.* **2011**, *15*, 4348–4356. [CrossRef]
41. Morris, G.Y.; Abbey, C.; Joos, G.; Marnay, C. A Framework for the Evaluation of the Cost and Benefits of Microgrids. In *CIGRE Int. Symposium*; 2011; pp. 1–14. Available online: https://www.osti.gov/servlets/purl/1050451/ (accessed on 15 September 2011).
42. Newman, B.Y.D. Right-sizing the grid. *Mech. Eng.* **2015**, *137*, 34–39. [CrossRef]
43. Rebours, Y.G.; Kirschen, D.S.; Trotignon, M. A Survey of Frequency and Voltage Control Ancillary Services—Part I: Technical Features. *IEEE Trans. Power System.* **2007**, *22*, 350–357. [CrossRef]
44. Rebours, Y.; Kirschen, D.S.; Trotingnon, M.; Rossignol, S. A Comprehensive Assessment of Markets for Frequency and Voltage control Ancillary Services. Ph.D. Thesis, CCSD, Las Vegas, NV, USA, 2008. Available online: https://tel.archives-ouvertes.fr/tel-00370805 (accessed on 25 March 2009).
45. Blanco, H.; Faaij, A. A review at the role of storage in energy systems with a focus on Power to Gas and long-term storage. *Renew. Sustain. Energy Rev.* **2018**, *81*, 1049–1086. [CrossRef]
46. Kirschen, D.; Strbac, G. Fundamentals of Power System Economics. John Wiley & Sons; 2004. Available online: http://www.cppa.gov.pk/DownloadFiles/Market%20Literature/Fundamentals%20of%20Power%20System%20Economics.pdf-181002094648397.pdf (accessed on 15 December 2020).
47. Švigir, N.; Kuzle, I.; Bosnjak, D. Ancillary Services in Deregulated Power Systems, WSEAS International conference on power systems. In Proceedings of the 8th WSEAS International Conference on POWER SYSTEMS (PS 2008), Santander, Cantabria, Spain, 23–25 September 2008.
48. Explanatory Memorandum: Introduction of Ancillary Services in India. 2015. Available online: http://www.cercind.gov.in/2015/draft_reg/Ancillary_Services.pdf (accessed on 15 December 2020).
49. Hirst, E.; Kirby, B. Allocating Costs of Ancillary Services: Contingency Reserves and Regulation. 2016. *ORNL/TM* **2003**, *152*. Available online: http://www.consultkirby.com/files/Tm2003-152_Allocate_Res_Reg_Cost.pdf (accessed on 15 December 2020).
50. Federal Register, United States of America. 2019; p. 20426. Available online: https://www.federalregister.gov/citation/81-FR-20426 (accessed on 4 July 2016).
51. Guide to Ancillary Services in the National. 2015. Available online: https://www.aemo.com.au/-/media/Files/PDF/Guide-to-Ancillary-Services-in-the-National-Electricity-Market.pdf (accessed on 15 December 2020).
52. Methods and Tools for Costing Ancillary Services, CIGRE Task Force. *Electra.* 2001. Available online: https://e-cigre.org/publication/ELT_196_7 (accessed on 15 December 2020).
53. Günter, N.; Marinopoulos, A. Energy storage for grid services and applications: Classification, market review, metrics, and methodology for evaluation of deployment cases. *J. Energy Storage* **2016**, *8*, 226–234. [CrossRef]
54. Load-Frequency Control and Perfromance, UCTE Operation Handbook. 2009, pp. 1–32. Available online: https://eepublicdownloads.blob.core.windows.net/public-cdn-container/clean-documents/pre2015/publications/ce/oh/appendix1_v19.pdf (accessed on 16 June 2004).

55. Shi, Q.; Li, F.; Hu, Q.; Wang, Z. Dynamic demand control for system frequency regulation: Concept review, algorithm comparison, and future vision. *Electr. Power Syst. Res.* **2018**, *154*, 75–87. [CrossRef]

56. Hernández, J.C.; Sanchez-sutil, F.; Vidal, P.G.; Rus-casas, C. Primary frequency control and dynamic grid support for vehicle-to-grid in transmission systems. *Electr. Power Energy Syst.* **2018**, *100*, 152–166. [CrossRef]

57. Hydroelectric Power, U.S. Department of Interior, Power Resources Office. 2005; pp. 1–26. Available online: https://www.usbr.gov/power/edu/pamphlet.pdf (accessed on 15 December 2020).

58. Izadkhast, S.; Garcia-gonzalez, P.; Frías, P. An Aggregate Model of Plug-In Electric Vehicles for Primary Frequency Control. *IEEE Trans. Power Syst.* **2020**, *30*, 1475–1482. [CrossRef]

59. Erik, E.; Michael, M.; Brendan, K. A Comprehensive Review of Current Strategies, Studies, and Fundamental Research on the Impact that Increased Penetration of Variable Renewable Generation has on Power System Operating Reserves. National Laboratory of the U.S. Department of Energy, Office of Energy Efficiency & Renewable Energy; 2011. Available online: https://www.nrel.gov/docs/fy11osti/51978.pdf (accessed on 15 December 2020).

60. Thien, T.; Schweer, D.; Moser, A.; Uwe, D. Real-world operating strategy and sensitivity analysis of frequency containment reserve provision with battery energy storage systems in the german market. *J. Energy Storage* **2017**, *13*, 143–163. [CrossRef]

61. Kumar, G.V.B.; Sarojini, R.K.; Palanisamy, K.; Padamnabhan, S.K.; Holm-nielsen, J.B. Large Scale Renewable Energy Integration: Issues and Solutions. *Energies* **2019**, *12*, 1996. [CrossRef]

62. Wang, X.; Wang, C.; Xu, T.; Guo, L.; Li, P.Y.; Meng, H. Optimal voltage regulation for distribution networks with multi-microgrids. *Appl. Energy* **2018**, *210*, 1027–1036. [CrossRef]

63. B. Operations, PJM Manual. 2019, p. 12. Available online: https://pjm.com/-/media/documents/manuals/m12-redline.ashx (accessed on 26 March 2020).

64. Mahzarnia, M.M.; Sheikholislami, A.; Adabi, J. A voltage stabilizer for a microgrid system with two types of distributed generation resources. *IIUM Eng. J.* **2013**, *14*, 191–205. [CrossRef]

65. Bevrani, H.; Ghosh, A.; Ledwich, G.F. Renewable energy sources and frequency regulation: Survey and new perspectives. *IET Renew. Power Gen.* 2010. [CrossRef]

66. Dehghanpour, K.; Afsharnia, S. Electrical demand side contribution to frequency control in power systems: A review on technical aspects. *Renew. Sustain. Energy Rev.* **2015**, *41*, 1267–1276. [CrossRef]

67. Ulbig, A.; Borsche, T.S.; Andersson, G. Impact of Low Rotational Inertia on Power System Stability and Operation. *IFAC Proc. Vol.* **2014**, *47*, 7290–7297. [CrossRef]

68. Díaz-gonzález, F.; Hau, M.; Sumper, A.; Gomis-bellmunt, O. Participation of wind power plants in system frequency control: Review of grid code requirements and control methods. *Renew. Sustain. Energy Rev.* **2014**, *34*, 551–564. [CrossRef]

69. Tielens, P.; Van Hertem, D. Receding Horizon Control of Wind Power to Provide Frequency Regulation. *IEEE Trans. Power Syst.* **2017**, *32*, 2663–2672. [CrossRef]

70. Yu, M.; Booth, C.D.; Roscoe, A.J. A Review of Control Methods for providing frequency response in VSC-HVDC transmission systems. In Proceedings of the 2014 49th International Universities Power Engineering Conference (UPEC), Cluj-Napoca, Romani, 2–5 September 2014; pp. 1–6. [CrossRef]

71. Revel, G.; Leon, A.E.; Alonso, D.M.; Moiola, J.L. Dynamics and Stability Analysis of a Power System with a PMSG-Based Wind Farm Performing Ancillary Services. *IEEE Trans. Circuits Syst. I Regul. Pap.* **2014**, *61*, 2182–2193. [CrossRef]

72. Dreidy, M.; Mokhlis, H.; Mekhilef, S. Inertia response and frequency control techniques for renewable energy sources: A review. *Renew. Sustain. Energy Rev.* **2017**, *69*, 144–155. [CrossRef]

73. Kanchanaharuthai, A.; Chankong, V.; Kenneth, A. Transient Stability and Voltage Regulation in Multimachine Power Systems Vis-à-Vis STATCOM and Battery Energy Storage. *IEEE Trans. Power Syst.* **2015**, *30*, 2404–2416. [CrossRef]

74. Justo, J.J.; Mwasilu, F.; Jung, J. Enhanced crowbarless FRT strategy for DFIG based wind turbines under three-phase voltage dip. *Electr. Power Syst. Res.* **2017**, *142*, 215–226. [CrossRef]

75. Ambati, B.B.; Kanjiya, P.; Khadkikar, V. A Low Component Count Series Voltage Compensation Scheme for DFIG WTs to Enhance Fault Ride-Through Capability. *IEEE Trans. Energy Convers.* **2015**, *30*, 208–217. [CrossRef]

76. Huang, P.H.; El Moursi, M.S.; Hasen, S.A. Novel Fault Ride-Through Scheme and Control Strategy for Doubly Fed Induction Generator-Based Wind Turbine. *IEEE Trans. Energy Convers.* **2015**, *30*, 635–645. [CrossRef]

77. Mendes, V.F.; De Sousa, C.V.; Rabelo, B.C.; Hofmann, W. Modeling and Ride-Through Control of Doubly Fed Induction Generators during Symmetrical Voltage Sags. *IEEE Trans. Energy Convers.* **2011**, *26*, 1161–1171. [CrossRef]

78. Hussein, A.A.; Ali, M.H. Comparison among series compensators for transient stability enhancement of doubly fed induction generator based variable speed wind turbines. *IET. Renew. Power Gen.* **2016**, *10*, 116–126. [CrossRef]

79. Guo, W.; Xiao, L.; Dai, S.; Xu, X.; Li, Y.; Wang, Y. Evaluation of the Performance of BTFCLs for Enhancing LVRT Capability of DFIG. *IEEE Trans. Power Electron.* **2015**, *30*, 3623–3637. [CrossRef]

80. Vidal, J.; Abad, G.; Arza, J.; Aurtenechea, S. Single-Phase DC Crowbar Topologies for Low Voltage Ride Through Fulfillment of High-Power Doubly Fed Induction Generator-Based Wind Turbines. *IEEE Trans. Energy Convers.* **2013**, *28*, 768–781. [CrossRef]

81. Huang, P.; Shawky, M.; Moursi, E.; Xiao, W.; Kirtley, L.K., Jr. Novel Fault Ride-Through Configuration and Transient Management Scheme for Doubly Fed Induction Generator. *IEEE Trans. Energy Convers.* **2013**, *28*, 86–94. [CrossRef]

82. Shen, Y.; Ke, D.P.; Sun, Y.Z.; Kirschen, D.S.; Qiao, W.; Deng, X.T. Advanced Auxiliary Control of an Energy Storage Device for Transient Voltage Support of a Doubly Fed Induction Generator. *IEEE Trans. Sustain. Energy* **2016**, *7*, 63–76. [CrossRef]

83. Yan, X.; Gu, C.; Zhang, X. Robust Optimization-Based Energy Storage Operation for System Congestion Management. *IEEE Syst. J.* **2019**, 1–9. [CrossRef]

84. Bai, L.; Wang, J.; Wang, C.; Chen, C.; Li, F. Distribution Locational Marginal Pricing (DLMP) for Congestion Management and Voltage Support. *IEEE Trans. Power Syst.* **2018**, *33*, 4061–4073. [CrossRef]

85. Crampes, C.; Trochet, J. Economics of stationary electricity storage with various charge and discharge durations. *J. Energy Storage* **2019**, *24*, 100746. [CrossRef]

86. Brahmendra Kumar, G.V.; Kumar, G.A.; Eswararao, S.; Gehlot, D. Modelling and Control of BESS for Solar Integration for PV Ramp Rate Contro. In Proceedings of the 2018 International Conference on Computation of Power, Energy, Information and Communication (ICCPEIC), Chennai, India, 28–29 March 2018; pp. 368–374.

87. Kalyani, M.K.; Vaidya, G.A. Ancillary Services through Microgrid for Grid Stabilty and Relaibility. 2017. Available online: https://www.electricalindia.in/ancillary-services-through-microgrid-for-grid-security-reliability/#:~{}:text=T%26D-,Ancillary%20services%20through%20Microgrid%20for%20Grid%20Security%20%26%20Reliability,of%20system%20and%20reduce%20congestion (accessed on 5 September 2017).

88. Zhang, S.; Tang, Y. Optimal schedule of grid-connected residential PV generation systems with battery storages under time-of-use and step tariffs. *J. Energy Storage* **2019**, *23*, 175–182. [CrossRef]

89. Bassett, K.; Carriveau, R.; Ting, D.S. Energy arbitrage and market opportunities for energy storage facilities in Ontario. *J. Energy Storage* **2018**, *20*, 478–484. [CrossRef]

90. Luo, X.; Wang, J.; Dooner, M.; Clarke, J. Overview of current development in electrical energy storage technologies and the application potential in power system operation. *Appl. Energy* **2014**, *137*, 511–536. [CrossRef]

91. Lavoine, O.; Regairaz, F.; Baker, T.; Belmans, R.; Meeus, L.; Vandezande, L.; Hewicker, C.; Matsubara, Y.; Pereira, R.; Torres, E.; et al. Ancillary Services: An Overview of International Practices. *Electra* **2010**, *252*, 86–91. Available online: http://hdl.handle.net/1814/40007. (accessed on 15 December 2020).

92. Kim, A.; Seo, H..; Kim, G.; Park, M.; Yu, I.; Otsuki, Y.; Tamura, J.; Kim, S.; Sim, K.; Seong, K. Operating Characteristic Analysis of HTS SMES for Frequency Stabilization of Dispersed Power Generation System. *IEEE Trans. Appl. Super Conduct.* **2010**, *20*, 1334–1338.

93. Tan, X.; Li, Q.; Wang, H. Advances and trends of energy storage technology in Microgrid. *Int. J. Electr. Power Energy Syst.* **2013**, *44*, 179–191. [CrossRef]

94. Ancillary Services Unbundling Electricity Products—An Emerging Market Eurelectric. 2004. Available online: http://pierrepinson.com/31761/Literature/Eurelectric2004-ancillaryservices.pdf (accessed on 15 December 2020).

95. FERC–Federal Energy Regulatory Commission. Promoting Wholesale Competition through Open-Access Non-Discriminatory Transmission Service by Public Utilities. 2005. Available online: https://www.ferc.gov/whats-new/comm-meet/091505/E-1.pdf (accessed on 24 May 1996).

96. Summary of Discussion Paper on Re-designing Ancillary Services Mechanism in India. 2018; pp. 1–51. Available online: cercind.gov.in/2018/draft_reg/DP.pdf (accessed on 15 December 2020).

97. Ming, Z.; Ximei, L.; Lilin, P. The ancillary services in China: An overview and key issues. *Renew. Sustain. Energy Rev.* **2014**, *36*, 83–90. [CrossRef]

98. Module 6: Ancillary Services, National Programme on Technology Enhanced Learning. Available online: http://150.107.117.36/NPTEL_DISK4/NPTEL_Contents/Web_courses/Phase2_web/108101005/ancillary%20service%20management/introduction.html. (accessed on 15 December 2020).

99. Palizban, O.; Kauhaniemi, K.; Guerrero, J.M. Microgrids in active network management—Part I: Hierarchical control, energy storage, virtual power plants, and market participation. *Renew. Sustain. Energy Rev.* **2014**, *36*, 428–439. [CrossRef]

100. Report, Ancillary Services. 2018, pp. 1–48. Available online: https://www.dena.de/fileadmin/dena/Publikationen/PDFs/2019/2018_Innovation_report_ancillary_services.pdf. (accessed on 15 December 2020).

101. Singh, G.; Dey, K.; Kumar, K.V.N.P.; Kumar, A.; Rehman, S.; Gaur, K. Ancillary Services in India-Evolution, Implementation and Benefits. In Proceedings of the 2016 National Power Systems Conference (NPSC), Bhubaneswar, India, 19–21 December 2016; pp. 1–6. [CrossRef]

102. Supercharged: Challenges and Opportunities in Global Battery Storage Markets. 2019, pp. 1–26. Available online: https://www2.deloitte.com/content/dam/Deloitte/bg/Documents/energy-resources/gx-er-challenges-opportunities-global-battery-storage-markets.pdf. (accessed on 15 December 2020).

103. Annual Electricity Report in France. 2015. Available online: https://www.agora-energiewende.de/fileadmin2/Projekte/2014/CP-Frankreich/CP_France_1015_update_web.pdf (accessed on 15 December 2020).

104. ABC and AGC Interface Requirements AEMO. 2018. Available online: https://www.aemo.com.au/-/media/files/electricity/wem/security_and_reliability/ancillary-services/2018/abc-and-agc-requirements-sept-2018.pdf?la=en&hash=DF420D332F1552755E73C8A258D962F0 (accessed on 15 December 2020).

105. Energy, U.K. Ancillary Services Report. 2017. Available online: https://www.energy-uk.org.uk/publication.html?task=file.download&id=6138 (accessed on 15 December 2020).

106. Vietor, R.H.K.; Thomson, H.S. Mexico's Energy Reform. 2017, pp. 1–32. Available online: https://www.hbs.edu/faculty/Pages/item.aspx?num=52187 (accessed on 23 January 2017).

107. Asgari, M.H.; Tabatabaei, M.J.; Riahi, R.; Mazhabjafari, A.; Mirzaee, M.; Bagheri, H.R. Establishment of regulation service market in iran restructured power system. *Can. Conf. Electr. Comput. Eng.* **2008**, 713–718. [CrossRef]

108. Energy Services Market Intelligence Report. 2018. Available online: https://www.greencape.co.za/assets/Uploads/GreenCape-Energy-Services-2018-MIR-25052019.pdf. (accessed on 15 December 2020).

109. Electricity Market Side Service Regulation Market, Turkey: Purpose, Scope, Basis and Definitions. 2017. Available online: https://www.lexology.com/library/detail.aspx?g=2577d4f0-9783-4346-951c-9f8ac552ffbb (accessed on 26 November 2017).

110. Mengelkamp, E.; Gärttner, J.; Rock, K.; Kessler, S.; Orsini, L. Designing microgrid energy markets A case study: The Brooklyn Microgrid. *Appl. Energy* **2018**, *210*, 870–880. [CrossRef]

111. Kumar, G.V.B.; Palanisamy, K. Interleaved Boost Converter for Renewable Energy Application with Energy Storage System. In Proceedings of the 2019 IEEE 1st International Conference on Energy, Systems and Information Processing (ICESIP), Chennai, India, 4–6 July 2019; pp. 1–5. [CrossRef]

112. Lo, C.; Hobbs, B.F. A cooperative game theoretic analysis of incentives for microgrids in regulated electricity markets. *Appl. Energy* **2016**, *169*, 524–541. [CrossRef]

113. Yue, J.; Hu, Z.; Anvari-moghaddam, A. A Multi-Market-Driven Approach to Energy Scheduling of Smart Microgrids in Distribution Networks. *Sustainability* **2019**, *11*, 301. [CrossRef]

114. Zhang, C.; Wu, J.; Zhou, Y.; Cheng, M.; Long, C. Peer-to-Peer energy trading in a Microgrid. *Appl. Energy* **2018**, *220*, 1–12. [CrossRef]

115. Soshinskaya, M.; Crijns-graus, W.H.J.; Guerrero, J.M.; Vasquez, J.C. Microgrids: Experiences, barriers and success factors. *Renew. Sustain. Energy Rev.* **2014**, *40*, 659–672. [CrossRef]

116. Meyer, D.; Glotfelty, J.W. U.S.-Canada Power System Outage Task Force, Energy Department. 2003. Available online: https://digital.library.unt.edu/ark:/67531/metadc26005/ (accessed on 14 August 2003).

117. Wang, W.; Lu, Z. Cyber security in the Smart Grid: Survey and challenges. *Comput. Netw.* **2018**, *57*, 1344–1371. [CrossRef]

118. Committee on the Societal and Economic Impacts of Severe Space Weather Events: A Workshop, National Research Council United States of America. 2008. Available online: http://lasp.colorado.edu/home/wpcontent/uploads/2011/07/lowres-Severe-Space-Weather-FINAL.pdf. (accessed on 15 December 2020).

119. Maize, K. The Great Solar Storm of 2012? *Power Mag.* 2011. Available online: http://www.powermag.com/the-great-solar-storm-of-2012 (accessed on 11 February 2011).

120. Foster, J.S.; Gjelde, E.; Graham, W.R.; Hermann, R.J.; Kluepfel, H.M.; Lawson, R.L.; Gordon, K.S.; Lowell, L.W., Jr.; Joan, B.W. Report of the Commission to Assess the Threat to the United States from Electromagnetic Pulse (EMP) Attack. Volume 1: Executive Report. DTIC Document. 2004. Available online: http://www.empcommission.org/docs/empc_exec_rpt.pdf (accessed on 15 December 2020).

121. Maize, K. EMP: The Biggest Unaddressed Threat to the Grid. *Power Mag.* 2013. Available online: http://www.powermag.com/emp-the-biggest-unaddressed-threat-to-the-grid (accessed on 1 July 2013).

122. Mikova, B.T. Cyber Attack on Ukrainian Power Grid. 2018. Available online: https://is.muni.cz/th/uok5b/BP_Mikova_final.pdf. (accessed on 15 December 2020).

123. Times of Israel. Steinitz: Israel's Electric Authority Hit by Severe Cyber-Attack. 2016. Available online: https://www.timesofisrael.com/steinitz-israels-electric-authority-hit-by-severe-cyber-attack/ (accessed on 26 January 2016).

124. Smith, R. Assault on California Power Station Raises Alarm on Potential for Terrorism. *Wall Street Journal.* 2014. Available online: https://www.wsj.com/articles/assault-on-california-power-station-raises-alarm-on-potential-for-terrorism-1391570879 (accessed on 5 February 2014).

125. Smith, R. How America Could Go Dark. *Wall Street Journal.* 2016. Available online: https://www.wsj.com/articles/how-america-could-go-dark-1468423254 (accessed on 14 July 2016).

126. Campbell, R.J. Weather-Related Power Outages and Electric System Resiliency, Congressional Research Service, Library of Congress. 2012. Available online: https://fas.org/sgp/crs/misc/R42696.pdf (accessed on 28 August 2012).

127. Tweed, K. Con Ed to Batteries, Microgrids and Efficiency to Delay $1B Substation Build. 2014. Available online: https://www.greentechmedia.com/articles/read/con-ed-looks-to-batteries-microgrids-and-efficiency-to-delay-1b-substation (accessed on 17 July 2014).

128. Lovins, A.B. Rocky Mountain Institute editors. In *Small Is Profitable*, 1st ed.; Rocky Mountain Institute: Snowmass, CO, USA, 2002; Available online: https://rmi.org/wp-content/uploads/2017/05/RMI_Document_Repository_Public-Reprts_U02-09_SmallIsProfitableBook.pdf. (accessed on 15 December 2020).

129. Asmus, P.; Larence, M. Emerging Microgrid Business Models. 2016. Available online: http://www.g20ys.org/upload/auto/abf2f0a71ea657d34c551214a4ff7045515582eb.pdf. (accessed on 15 December 2020).

130. Guerrero, J.M.; Loh, P.C.; Lee, T.; Chandorkar, M. Advanced Control Architectures for Intelligent Microgrids—Part II: Power Quality, Energy Storage, and AC/DC Microgrids. *IEEE Trans. Ind. Electron.* **2013**, *60*, 1263–1270. [CrossRef]

131. Katiraei, F.; Iravani, R.; Hatziargyriou, N.; Dimeas, A. Microgrids Management–Control and Operation Aspects of Microgrids. *IEEE Power Energy Mag.* **2008**, *6*, 54–65. [CrossRef]

132. Lopes, J.A.P.; Madureira, A.G.; Moreiran, C.C.L.M. A view of microgrids. *Wiley Interdiscip. Rev. Energy Environ.* **2013**, *2*, 86–103. [CrossRef]

133. Bhatnagar, D.; Currier, A.; Hernandez, J.; Ma, O.; Kirby, B. Market and Policy Barriers to Energy Storage Deployment. Sandia National Laboratories/Office of Energy Efficiency and Renewable Energy; 2013. Available online: https://www.sandia.gov/ess-ssl/publications/SAND2013-7606.pdf (accessed on 15 December 2020).

134. Byrne, R.H.; Concepcion, R.J.; Silva-monroy, A. Estimating Potential Revenue from Electrical Energy Storage in PJM. 2016; pp. 1–5. Available online: https://www.osti.gov/servlets/purl/1239334 (accessed on 1 February 2016).

135. Denholm, P.; O'Connell, M.; Brinkman, G.; Jorgenson, J. Overgeneration from Solar Energy in California: A Field Guide to the Duck Chart, Natl. Renew Energy Lab Tech rep. NRELTP-6A20-65023; 2015. Available online: https://www.nrel.gov/docs/fy16osti/65023.pdf (accessed on 15 December 2020).

136. California Independent System Operator. What the Duck Curve Tells Us about Managing a Green Grid. 2016. Available online: https://www.caiso.com/Documents/FlexibleResourcesHelpRenewables_FastFacts.pdf (accessed on 15 December 2020).

137. Bebic, J. *Power System Planning: Emerging Practices Suitable for Evaluating the Impact of High-Penetration Photovoltaics*; National Renewable Energy Laboratory: Golden, CO, USA, 2008. Available online: https://www.nrel.gov/docs/fy08osti/42297.pdf (accessed on 15 December 2020).

138. Hoke, A.; Butler, R.; Hambrick, J.; Kroposki, B. *Maximum Photovoltaic Penetration Levels on Typical Distribution Feeders*; Natl. Renew Energy Lab.: Golden, CO, USA, 2012. Available online: https://www.nrel.gov/docs/fy12osti/55094.pdf (accessed on 15 December 2020).

139. Martin, R. Texas and California Have a Bizarre Problem: Too much Renewable Energy. MIT Technol Rev. 2016. Available online: https://www.technologyreview.com/2016/04/07/161139/texas-and-california-have-too-much-renewable-energy/ (accessed on 7 April 2016).

140. Martin, R. Loading up on Wind and Solar Is Causing New Problems for Germany, MIT Technol Rev. 2016. Available online: https://www.technologyreview.com/2016/05/24/159991/germany-runs-up-against-the-limits-of-renewables/ (accessed on 24 May 2016).

141. Perez, E. Investigators Find Proof of Cyber-Attack on Ukraine Power Grid. CNN. 2016. Available online: https://edition.cnn.com/2016/02/03/politics/cyberattack-ukraine-power-grid/index.html#:~{}:text=Washington%20 (accessed on 4 February 2016).

142. Akhil, A.A.; Huff, G.; Aileen, B.; Currier, B.; Benzamin, C.K.; Rastler, D.M.; Chen, S.B.; Cotter, A.L.; Bradshaw, D.T.; Gauntlett, W.D.; et al. *DOE/EPRI 2013 Electricity Storage Handbook in Collaboration with NRECA*; Albuquerque, N.M., Ed.; Sandia National Laboratories: Albuquerque, NM, USA, 2013. Available online: https://prod-ng.sandia.gov/techlib-noauth/access-control.cgi/2015/151002.pdf (accessed on 15 December 2020).

143. Eyer, J.; Corey, G. *Energy Storage for the Electricity Grid: Benefits and Market Potential Assessment Guide, Sandia National Laboratories Report*; SAND2010-0815; Sandia National Laboratories: Albuquerque, NM, USA, 2010. [CrossRef]

144. Kumar, G.V.B.; Kaliannan, P.; Padmanaban, S.; Holm-Nielsen, J.B.; Blaabjerg, F. Effective Management System for Solar PV Using Real-Time Data with Hybrid Energy Storage System. *Appl. Sci.* **2020**, *10*, 1108. [CrossRef]

145. Laaksonen, H.; Kauhaniemi, K. Control Principles for Black Start and Island Operation of Microgrid. Nordic Workshop on Power and Industrial Electronics (NORPIE/). 2008. Available online: https://aaltodoc.aalto.fi/handle/123456789/810. (accessed on 15 December 2020).

146. Lopes, J.A.P.; Moreira, C.L.; Resende, F.O. Microgrids Black Start and Islanded Operation. 2005, pp. 22–26. Available online: http://www.montefiore.ulg.ac.be/services/stochastic/pscc05/papers/fp69.pdf (accessed on 26 August 2005).

147. Xu, G.; Xu, L.; Morrow, J. Power oscillation damping using wind turbines with energy storage systems. *IET Renew. Power Gener.* **2013**, *7*, 449–457. [CrossRef]

148. Beza, M.; Bongiorno, M. Power Oscillation Damping Controller by Static Synchronous Compensator with Energy Storage. In Proceedings of the IEEE Energy Conversion Congress and Exposition, Phoenix, AZ, USA, 17–22 September 2011; pp. 2977–2984.

149. Altin, M.; Teodorescu, R.; Jensen, B.B.; Annakkage, U.D.; Iov, F.; Kjaer, P.C. Methodology for Assessment of Inertial Response from Wind Power Plants. In Proceedings of the 2012 IEEE Power and Energy Society General Meeting, San Diego, CA, USA, 22–26 July 2012; pp. 1–8.

150. Teodorescu, R.; Rodriguez, P. Lifetime Investigations of a Lithium Iron Phosphate (lip) Battery System Connected to a Wind Turbine for Forecast Improvement and Output Power Gradient Reduction. *Battcon Arch. Pap.* **2012**, 1–8.

151. Chua, K.H.; Lim, Y.S.; Wong, J.; Taylor, P.; Morris, E.; Morris, S. Voltage Unbalance Mitigation in Low Voltage Distribution Networks with Photovoltaic Systems. *J. Electron. Sci. Technol.* **2012**, *10*, 1–6. [CrossRef]

152. Levron, Y.; Shmilovitz, D. Power systems optimal peak-shaving applying secondary storage. *Electr. Power Syst. Res.* **2012**, *89*, 80–84. [CrossRef]

153. Callaway, D.S. Tapping the energy storage potential in electric loads to deliver load following and regulation with application to wind energy. *Energy Convers. Manag.* **2009**, *50*, 1389–1400. [CrossRef]

154. Thounthong, P.; Rael, S.; Davat, B. Analysis of Supercapacitor as Second Source Based on Fuel Cell Power Generation. *IEEE Trans. Energy Convers.* **2009**, *24*, 247–255. [CrossRef]

155. Kim, J.; Jeon, J.; Kim, S.; Cho, C.; Park, J.H.; Kim, H.; Nam, K. Cooperative Control Strategy of Energy Storage System and Microsources for Stabilizing the Microgrid during Islanded Operation. *IEEE Trans. Power Electronics.* **2010**, *25*, 3037–3048.
156. Mercier, P.; Cherkaoui, R.; Oudalov, A. Optimizing a Battery Energy Storage System for Frequency Control Application in an Isolated Power System. *IEEE Trans. Power Syst.* **2009**, *24*, 1469–1477. [CrossRef]

Publisher's Note: MDPI stays neutral with regard to jurisdictional claims in published maps and institutional affiliations.

 *inventions*

Article

Secondary Voltage Control Application in a Smart Grid with 100% Renewables

Omar H. Abdalla [1], Hady H. Fayek [2,*] and A. M. Abdel Ghany [1]

[1] Electrical Power and Machines Dept., Faculty of Engineering, Helwan University, Cairo 11792, Egypt; omar.hanafy@h-eng.helwan.edu.eg (O.H.A.); ghanyghany@hotmail.com (A.M.A.G.)
[2] Electromechanics Engineering Dept., Faculty of Engineering, Heliopolis University, Cairo 11785, Egypt
* Correspondence: hadyhabib@hotmail.com; Tel.: +20-1005472291

Received: 16 June 2020; Accepted: 30 July 2020; Published: 1 August 2020

Abstract: This paper presents secondary voltage control by extracting reactive power from renewable power technologies to control load buses voltage in a power system at different operating conditions. The study is performed on a 100% renewable 14-bus system. Active and reactive powers controls are considered based on grid codes of countries with high penetration levels of renewable energy technologies. A pilot bus is selected in order to implement the secondary voltage control. The selection is based on short-circuit calculation and sensitivity analysis. An optimal Proportional Integral Derivative (PID) voltage controller is designed using genetic algorithm. A comparison between system with and without secondary voltage control is presented in terms of voltage profile and total power losses. The optimal voltage magnitudes at busbars are calculated to achieve minimum power losses using optimal power flow. The optimal placement of Phasor Measurement Units (PMUs) is performed in order to measure the voltage magnitude of buses with minimum cost. Optimization and simulation processes are performed using DIgSILENT and MATLAB software applications.

Keywords: 100% renewable power system; secondary voltage control; tertiary voltage control; grid code; wind farms; photovoltaic parks

---

1. Introduction

One of the main visions of the future power systems is high penetration level of renewable energy technologies [1,2]. Renewable energy power plants, such as wind and solar farms, are fundamentally different from conventional power plants in terms of components and structure. In conventional power systems, synchronous generators are the main sources of reactive power to support voltage and maintain stability. In the renewable generation system, the voltage is controlled through power electronic converter [3–5].

If the grid faces voltage instability, following a disturbance, the voltage declines within few minutes dramatically and monotonically. When this decrease is too pronounced, the system integrity is endangered. This degradation process may eventually lead to a blackout in the form of a voltage collapse [6]. The control of voltage in electrical power networks has three hierarchical levels: Primary Voltage Control (PriVC), Secondary Voltage Control (SecVC), and Tertiary Voltage Control (TerVC) [7]. PriVC regulates the voltage of the generator bus by controlling the reactive power injection or absorption through automatic voltage regulators for the synchronous generators [8]. PriVC is performed in fraction of seconds to ten seconds. SecVC is directed to regulate load buses voltage magnitude. SecVC requires partitioning the grid into regions and select the pilot bus of each region. The pilot bus is normally the most sensitive load bus to the reactive power changes. A control system is applied to the pilot bus of the region by using a controller to track the optimal value of the pilot bus voltage. The optimal value is calculated from TerVC. Normally, through control action the generators in the region will be asked to inject/absorb reactive power by adding a signal to the PriVC. SecVC implementation time is ranging from 50 s to 25 min. to avoid overlap with the PriVC action. TerVC is normally an optimal power flow process occur every half an hour or longer time interval. The optimal power flow process is typically performed to achieve an objective or multi objectives, such as minimum power losses or/and minimum generating cost. Pilot bus reference voltage is calculated from TerVC [9].

Countries worldwide are spotting remarkable efforts towards the smart grid in the recent years. The smart grid is a power system with high quality, security, stability, efficiency, and green environment. To achieve these objectives wide area measurement systems, real time control, real time protection, and self-healing would be considered in the future grids. The wide area measurement system is achieved by using Phasor Measurement Units (PMUs) based on global positioning system. PMUs facilitate real time voltage control and self-healing [10,11]. An application of coordinated SecVC to a power system with conventional power plants was presented in [12].

There are a lot of methods proposed in previous work in order to apply primary voltage control to transmission and distribution grids with high share of renewable energies. In [13], the paper illustrates how the voltage control is applied through optimal placement of unified power flow controllers. In [14], the research improved the voltage performance of the grid through optimal placement and the sizing of renewable energies.

In [15], the paper illustrates a technique to apply distributed secondary voltage control among inverter-based generation systems. The secondary control is applied based on droop control. The applied method in [15]; leads to steady-state error in the pilot bus voltage due to absence of integral action in the regional controller. In addition, technical requirements of grid codes for integrating renewable energies have not been considered. In [16], the authors target secondary voltage applications of a 100% inverter-based microgrid showing the effect of communication delays. In [15] and [16], inverter based secondary voltage control is distributed and controlling buses, which include distributed generation.

The main contributions of this paper are:

- Coordinated Secondary voltage control is applied to control a load bus (not include a distributed generation) in a grid feeding from 100% renewable energies through extracting reactive power from inverter-based renewable energy power plants. In this research, a number of Proportional Integral Derivative (PID) controllers are designed in order to deal with various operating conditions using the genetic algorithm toolbox in MATLAB to enable the pilot bus to reach its optimal value at each condition.
- The grid codes of integrating renewable energies for countries with high share of renewable energies mainly photovoltaics (PV) and wind are considered while applying secondary voltage control and primary frequency control.
- PMUs are used to measure the real time voltage of the pilot bus and detect through what-if-analysis the optimal parameters of the secondary voltage genetic PID controller while using the neural network.

The paper is organized, as follows: Section 2 describes the configuration of the studied 100% renewable power system. Section 3 discusses grid code for renewables integration. Section 4 presents the secondary voltage control and pilot bus selection. Section 5 illustrates how the PMUs are optimally allocated. Section 6 briefly describes how the PID controller is optimally designed. Section 7 illustrates how tertiary voltage control is performed. In Section 8, the application of the coordinated SecVC to the 100% renewable power system is presented. Section 9 presents the simulation results and Section 10 summarizes the main conclusions.

2. 100% Renewable Grid

Some countries achieved 100% renewables in grid operation, such as Iceland, while Norway achieved 98% penetration level of renewables in 2019, Kenya achieved 70 % and aims to reach 100% by the end of 2020 while Brazil and Canada achieved more than 70% and 60%, respectively. Countries, like Denmark, Ireland and Germany, are operating their grids with more than 20% penetration level of renewables [17,18]. Now, it is a trend in the United States to reach 100% renewables through building new wind farms, PV parks, and distributed generation [19,20]. Globally, the current penetration level of renewables is 19.3% growing faster than the demand, which is an indicator to global 100% renewable energy within decades, which is also a goal of the European Commission [18].

In this paper, selecting a 100% renewable power system was an objective to ensure the validity of controlling the pilot bus load voltage control from only renewable generators. The power system chosen shown in Figure 1 is the IEEE 14-Bus system after modifying the generation to be 100% renewable, as listed in Table 1. The modifications are made to enable the control of the pilot bus from inverter-based generation systems. In the proposed 14 bus system, unlike countries with high participation of renewable energies, it is assumed that the penetration level of inverter-based power plants is higher than that of the hydro. The assumption is performed based on the global trend towards increasing the installed capacity of photovoltaics and wind generation. Appendix A illustrates the lines and load data.

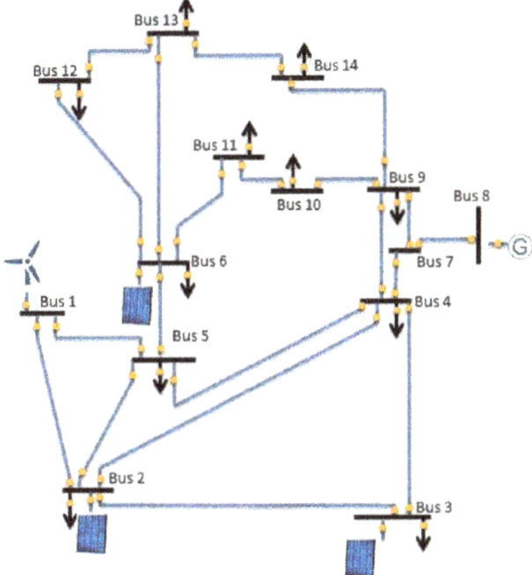

Figure 1. Bus system with 100% renewables.

Table 1. Generation systems in the 100% Renewable 14 bus system.

Power Station Type	Location Bus No.	Installed Capacity
DFIG wind farm	1	190 MW
Photovoltaic park	2	95 MW
Photovoltaic park	6	40 MW
Photovoltaic park	3	150 MW
Hydro power plant	8	140 MW

3. Renewable Energy Grid Integration

3.1. Wind Farm

There are four main types of wind turbines generators (WTGs). Type 1 and 2 WTGs are based on induction generators, and they cannot provide reactive power, so no chance to offer voltage control to the power system. Type 3 (with a doubly fed induction generator) and Type 4 (with a synchronous generator) WTGs both integrated to the grid through power electronic converters and DC link capacitor. For both types, the converters and the DC link offer the ability to provide reactive power and, hence, apply voltage regulation. In this paper the Type 3, which is shown in Figure 2, was selected to simulate the wind generators in the 100% renewable grid.

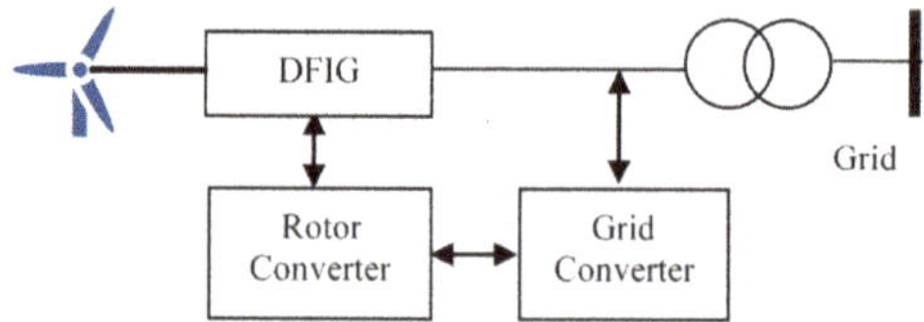

Figure 2. Wind energy generation system (Type 3) [18].

The Nordic grid code for integrating wind farms to the grid was taken in this study as a reference to establish the 100% 14 bus renewable grid based on the frequency active power control and voltage reactive power control, as explained in the following [21]:

3.1.1. Frequency and Active Power

Wind farms ≥ 10 MVA shall be able for instantaneous contribution of active reserve both for up- and down-regulation. Wind farms ≥ 10 MVA shall have the availability to send frequency-controlled reserve power, activated via frequency controllers implemented in each wind turbine. It shall be possible to set the droop from 1 to 12 % for wind farms ≥ 1 MVA [21].

3.1.2. Voltage and Reactive Power

Wind power plants shall be able to operate with a power factor limits 0.95 lagging and 0.95 leading, at nominal active power generation, as illustrated in Figure 3. The requirements are referred to the Point of Common Coupling (PCC) with the grid. The compensation shall be dimensioned according to the needs of the grid where the wind turbines are connected. It shall be possible to change the set-point for the voltage controller both locally and remotely [21].

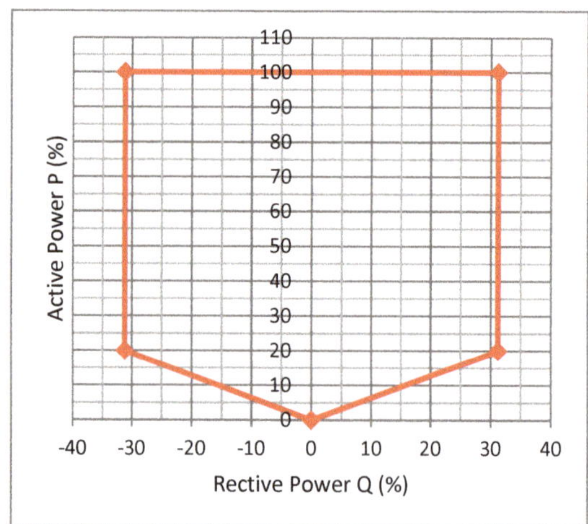

Figure 3. P-Q capability chart for renewable power plants [21].

WTGs should be capable of regulating voltage through the provision of dynamic reactive power support. However, wind farms are comprised of many distributed wind turbine generators and, therefore, exhibit behavior that is different from that of conventional high rating generators. Nevertheless, from a smart grid operational point-of-view, wind farms should provide voltage controllability that is required for system stability and reliability.

3.2. Photovoltaic Park

In recent years, photovoltaic systems have been widely used as distributed electric power generators in kW scale and sometimes as farms producing output power in range of MW scale. Despite the fact that the configuration of the photovoltaic farm is different than that of the traditional power plants. PV parks are required to support the grid in terms of active and reactive powers targeting stable and reliable system. Figure 4 shows the main components of a grid tied photovoltaic power plant. Active and reactive power control systems mainly depend on the change of the control signal of the two converters of the system.

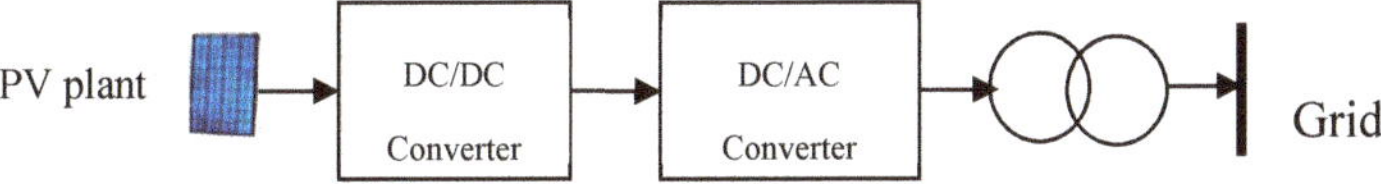

Figure 4. Grid tied PV system [18].

Based on some European countries grid code and Chinese grid code for integrating the PV to the grid, the following requirements should be applied [22,23]:

3.2.1. Frequency and Active Power Control

Grid-connected photovoltaic power stations are required to respond to the frequency deviation that is based on different grid codes. Different output response requirements are generally specified to three frequency ranges. In the range of e.g., ±0.2 Hz maximum deviation from the base frequency a dead band is set. The system frequency within this range needs no response of the photovoltaic park output. When the power system frequency is higher than 50.2 Hz, a PV park should reduce its active power; the control in this case is performed while using the frequency droop method. If the frequency is below 49.8 Hz, then active power control may also be applied. In existing codes of some countries, no active power control is required in this frequency range. The dead band in ENTSO-E and Germany codes is ±0.5 Hz [23].

3.2.2. Voltage and Reactive Power Control

The grid connected inverters installed in photovoltaic power station should meet the requirement that their power factors could be dynamically adjusted within the range of 0.95 leading 0.95 lagging at the rated active power output, as shown in Figure 3. Solar codes in some countries require reactive power control within the above range even at zero active power output [24].

If a fault occurs; to get faster grid voltage recovery, the grid-connected photovoltaic park is required to increase its reactive current output when the voltage dip occurs at PCC with the grid to provide voltage support. If the voltage at the PCC ranges between 0.9 p.u. to 1.1 p.u., no reactive current injection or absorption is enabled based on the grid code of many countries, but, below 0.9 p.u. a reactive current should be injected to the grid as fast as possible (less than 30 ms in many countries). If the voltage is higher than 1.1 p.u., reactive power absorption is required by the PV park.

3.3. Implementation of Power Frequency Control

Figure 5 shows a frequency control system of inverters-based generators in power systems. The three blocks are power block, current block, and voltage block. The current and voltage blocks are classified as fast dynamics blocks, while power block concerns with power control ignoring fast dynamics of the other two blocks. The basic key in the power controller is the droop control, since the inertia is small in most wind turbines and negligible in PV modules so the primary frequency control is made mainly using the droop control. Equation (1) describes the power frequency control [25].

$$f_i = f_0 - P_i n_{p_i} \tag{1}$$

where f_i is the output frequency while f_o is the desired frequency, P_i is the park or farm measured real power, and n_{p_i} is the frequency droop.

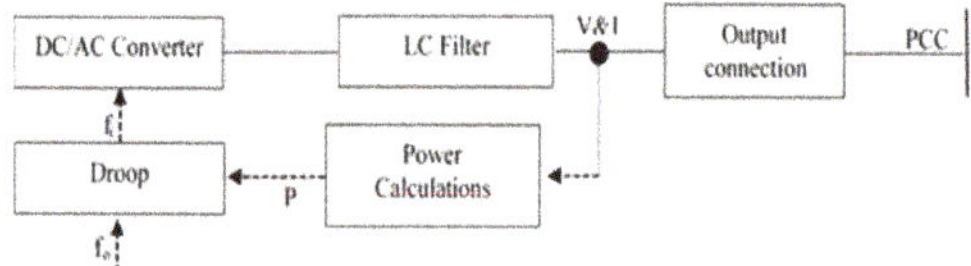

Figure 5. Inverter control blocks [25].

4. Voltage Control & Pilot Bus Selection

4.1. Secondary Voltage Control

To apply SecVC in a large electrical power network, the network is normally partitioned into regions while small networks are considered as a single region power system, as illustrated in [8]. The reactive power compensators/generators are responsible for absorbing or generating adequate reactive power to reach an optimal value. The SecVC main objective when applied to a smart grid consisting of renewable generators is to control the pilot bus voltages magnitude by changing the reference points of the firing circuits of the wind or photovoltaic inverters in a coordinated way.

Figure 6 illustrates how voltage control hierarchy is applied based renewable power generators in the electrical network with SecVC configuration details. The execution time of each level is less than the higher one in order to avoid overlapping between the three levels. This application can be implemented by adapting the dominant time constant of the SecVC to be higher than that of all the PriVCs in the power grid and the TerVC is performed every half an hour or longer time interval [9]. The voltage control sequence is performed, as described by the following procedure:

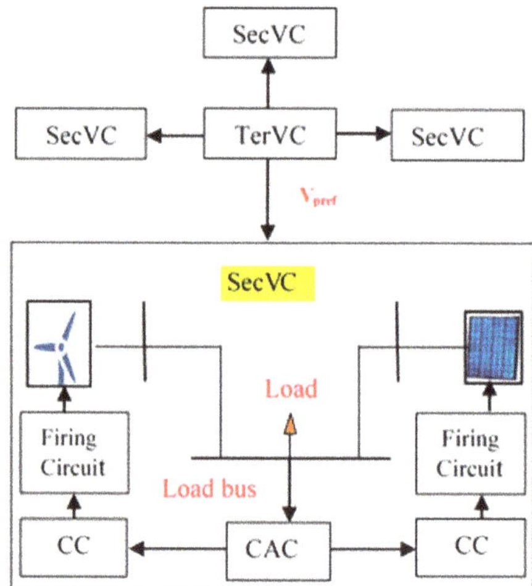

Figure 6. Voltage control strategy in 100% renewable grid.

PriVC (firing angle control) controls the PCC voltage magnitude of the farm or the park, which is implemented by controlling automatically the on–off switching of the inverter through controlling I_{qref} reference current, as shown in (2). The time constant of the PriVC is normally short (≤ 0.5 s). The actual measured I_{qmeas} is compared with the reference I_{qref}, as shown in (3), and the error (ΔI_q) is the input of the current regulator, which by its role will control V_d and V_q leading to the change of the firing angle of the inverter.

$$I_{qref} = K_v \int_0^t \left(V_{ref} - V\right)dt \tag{2}$$

$$\Delta I_q = I_{qref} - I_{qmeas} \tag{3}$$

Figure 7 shows Unit Cluster Control (CC). The PriVC voltage reference V_{ref} could be varied between V_{min} and V_{max}, which obtains the unit reactive power production Q_G corresponding to its reference value Q_{ref}, as illustrated in (4).

$$V_{ref} = \left| K_{IQ} \int_0^t (q - Q_G)dt \right|_{V_{min}}^{V_{max}} \tag{4}$$

where $K_{IQ} = \left(X_{TG} + X_{eq}\right)/T_G$ represents the regulator integral gain [6]. T_G, X_{TG}, and X_{eq} are dominant time constant, transformer power station reactance, and equivalent reactance of the line connecting the pilot bus with reactive power source, respectively.

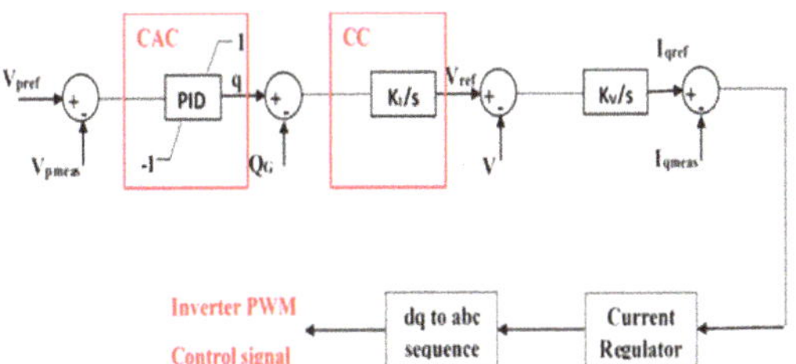

Figure 7. Renewable power plant SecVC scheme.

The pilot Central Area Control (CAC) provides reactive power level q. The reactive level q has minimum and maximum limits, which are qmin = −100% and its maximum qmax = 100% to track pilot bus reference value Vpref [9].

$$\Delta V_p = V_{pref} - V_{pmeas} \tag{5}$$

$$q = \left| K_p \Delta V_p + K_I \int_0^t \Delta V_p dt + K_D \frac{d\Delta V_p}{dt} \right|_{-1}^{+1} \tag{6}$$

where K_p, K_I, and K_D are the coordinated PID controller parameters.

V_{pref} can be obtained by performing optimal power flow calculation in order to achieve certain objective of operation when considering equality and non-equality constraints.

4.2. Pilot Bus Selection

The pilot bus is selected based on the idea of controlling this bus to control the voltages of all load buses in the voltage control area. The load bus with the highest voltage sensitivity $|\frac{\partial V}{\partial Q}|$ in each area is selected to be the pilot bus [7]. Short-circuit analysis is performed to define the load bus with highest short-circuit current to support this selection.

5. Phasor Measurement Units (PMUs)

PMU is a device that can measure voltage and current phasors in the transmission in real time. PMU is one of the most promising devices in smart grid, as it facilitates the application of many features such as real time control and protection in addition to self-healing. The distinction comes from its unique ability to provide synchronized phasor measurements of voltages and currents from widely dispersed locations in an electric power grid. The commercialization of the global positioning satellite (GPS) with an accuracy of timing pulses in the order of one microsecond made possible the commercial production of phasor measurement units [26]. The main question, which was raised in the past decade, is it possible to install a PMU in each bus bar? Because of the cost of the measurement devices, it is not economically feasible to place them at every grid bus, the answer is no [27]. In this paper, a sequential linear programming optimization technique is used to get real time measurements for the load buses voltages using a minimum number of PMUs [28].

The optimization problem is defined as follows:

- Objective function: minimization of the PMU devices in the power system,

$$Min \sum_{i}^{n} w_i x_i \tag{7}$$

Such that n is the total number of buses, w is 0 or 1 depending on the PMU capability of measurement and x is the bus number.

- Variables: buses that include PMU devices.
- Constraints: each bus voltage can be reached by at least one PMU, i.e.,

$$f(x) \geq 1 \tag{8}$$

f(x) is a vector function, whose entries are non- zero if the corresponding bus voltage is solvable while using the given measurements [27,28].

The PMU can measure the entire bus voltage and the voltage of other buses connected to the bus which has PMU. The optimal placement of PMUs is applied to the network that is shown in Figure 1. Performing the optimization on MATLAB, results in placing PMUs at buses 2, 6, 8, and 9.

6. PID Controller Design

The design of PID controllers will be performed for each operation condition separately using the GA toolbox in MATLAB to enable the pilot bus to reach its optimal value at this condition. The optimization problem is described, as follows:

- Objective Function: Minimizing the integration of pilot bus voltage error (ΔV_p):

$$Min \int_{0}^{t} \left(\Delta V_p\right)^2 dt \tag{9}$$

- Variables: coordinated PID controller parameters (K_p, K_I, and K_D).
- Constraints: q limits (±100%) and generated reactive power limits.

GA is widely known for its optimization capability results; it includes three main stages, which are: reproduction, crossover, and mutation. GA in each stage produce a new generation from the old one by selecting individuals. The convergence speed is varied by using different probabilities for applying these operators. Crossover and mutation operators must be designed carefully, because their selection highly contributes in the evaluation of genetic algorithm [29].

In the reproduction stage, the performance of individuals is measured by the fitness function, and it also directs the selection process. Suitable individuals increase opportunities to transfer genetically important material to successive generations. Through this route, GA searches in the search space from many points simultaneously, and the search focus is constantly narrowed to the observed performance areas [29]. The reproduction is responsible for selecting a novel form of chromosomes, and cross-over mix exchanging partitions of two chromosomes. More chromosomes are generated with the cross-over operation. Reproduction is a clear characteristic of existing species with significant reproductive potential that their population size will increase exponentially if individuals of this species proliferate successfully. Reproduction can progeny by the transfer of the individual's genetic program. The search space is widened to all possible groups of the parameter values of the controller in order to minimize the values of the fitness function, which is the error criterion. The error criterion of PID controller is minimized by GA in each iteration. In this paper, integral of square error is selected to be minimized as a fitness function for the design of optimal controllers [29]. The GA applied in this research has the following settings: population type: double vector, population size: 20. The Elite count reproduction: 2 and the crossover fraction: 0.8. To avoid premature convergence to local optima, the best set of parameters for the optimization problem are selected by adapting the number of population members and number of generations in addition to increasing the mutation rate in GA toolbox in MATLAB.

7. Tertiary Voltage Control

Tertiary Voltage Control (TerVC) is responsible for changing the setting point of the pilot bus voltages in order to implement Secondary Voltage Control (SecVC) based on optimal load flow process. Figure 6 shows the voltage control hierarchy. The optimal load flow is applied based on the following description:

- Objective function: minimization of total active power losses.
- Variables: buses voltage (magnitude and angle) and each renewable park/farm active and reactive power
- Constraints: the optimization process is subjected to the following conditions:

Load flow equations

$$\sum PG - \sum PD - \sum PLoss = 0 \tag{10}$$

$$\sum QG - \sum QD - \sum QLoss = 0 \tag{11}$$

Generating limits for each renewable park/farm

$$PG_{min} \leq PG \leq PG_{max} \tag{12}$$

$$QG_{min} \leq QG \leq QG_{max} \tag{13}$$

Bus voltage magnitude level limits

$$Vb_{min} \leq Vb \leq Vb_{max} \tag{14}$$

Each line loading thermal limit

$$Pline \leq Pline_{max} \tag{15}$$

In (14), the maximum and minimum voltage values differ from one operating condition to another. During normal operation, the minimum and maximum limits should be 95 % and 105% of the rated bus voltage magnitude, respectively, while during contingency (generator outage or line outage) the limits should be 90% and 110% of the rated bus voltage magnitude, respectively.

8. Proposed Control Strategy

The idea in this paper is that in high penetration level of renewables, the renewable power plants should support the voltage of loads at PQ buses in case of contingencies. In this research, we consider a 100% renewable 14-bus system with a very high penetration level of wind and photovoltaic generation, as shown in Figure 1. The voltage control strategy explained in Figure 8 includes two stages: the design stage, which is basically made to enable the grid to optimally behave during any operation stage through optimization process and artificial intelligence. During the operation stage, the system will be able to detect its entire case the optimal values of this case (certain contingency) and the controllers should work properly to reach those optimal values. The design stage flow chart shows that at different conditions of the power system optimal values change. In this study a renewable plant outage will be assumed, and the power system optimization process will take place in order to minimize the total power loss taking into considerations voltages, active, reactive power limits. The optimization will be performed using the DIgSILENT power factory software.

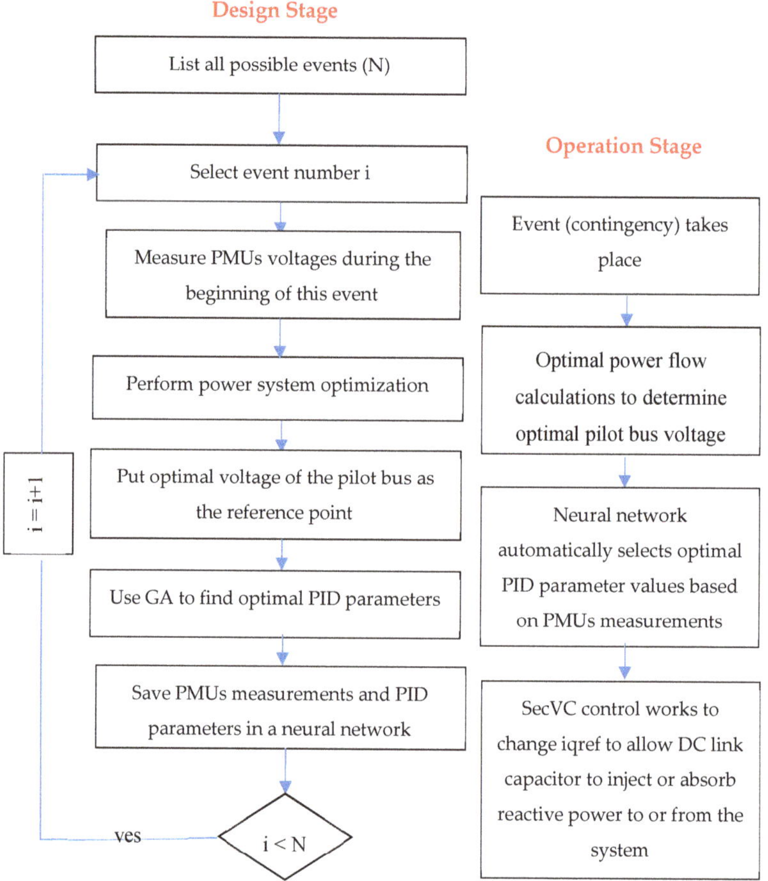

Figure 8. Flow chart of proposed voltage control strategy during design stage (**left**) and operation stage (**right**).

The secondary voltage control scheme, which is shown in Figure 7, includes parameters that are driven from the power system directly except the coordinated PID controller parameters, which is designed to track the reference value of the pilot load bus at a time range longer than that of the PriVC. The 100% renewable power network is subjected to different large disturbance, such as the outage of part of a renewable power plant or an outage of a transmission line. In each disturbance, the optimal parameters of the coordinated PID controller will differ due to the change of the power system configuration at each operating condition; moreover, the pilot bus voltage reference will change based on TerVC at each operating case. To solve this problem, an artificial intelligence element is used. The element is chosen to be the neural network to work based on what-if analysis such that it will be able to optimally tune the coordinated PID based on the operating condition. The neural network will select optimal PID controller parameters according to the operating condition that is determined by PMUs readings as shown in Figure 9.

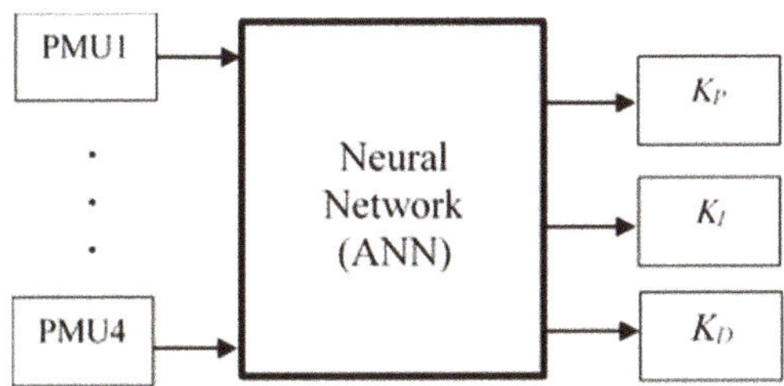

Figure 9. Neural network used in SecVC.

The neural network is designed as follows in MATLAB/SIMULINK:

- No. of inputs: four (since the network has 4 PMUs)
- No. of outputs: three (coordinated PID controller parameters)
- Hidden layer: one hidden layer with 10 neurons
- Training method: Lavenberg–Marquardt back propagation method

To beat the curse of dimensionality of ANN, all of the possible contingencies are considered, and optimal controllers are created based on genetic algorithm for each contingency.

9. Simulation Results

9.1. Pilot Bus Selection

Both sensitivity $|\partial V/\partial Q|$ and short-circuit techniques provide the same pilot bus result, which is bus 5, as shown in Table 2. It is the most sensitive bus with highest maximum short circuit current [7]. DIgSILENT software is used in order to calculate both the sensitivity and short circuit currents at all buses. The pilot bus is selected from the load buses assuming the 14-bus system consists of one voltage control area.

Table 2. Buses Sensitivity $|\partial V/\partial Q|$ and Maximum Short Circuit Current.

Bus No.	Bus Type	$\|\partial V/\partial Q\|$ PU/MVAR	Short Circuit Current in PU
1	Generator	0	34.9
2	Generator	0	20.99
3	Generator	0	11.93
4	Load	0.001286	14.81
5	Load	0.001967	15.19
6	Generator	0	10.78
7	Load	0.000939	8.64
8	Generator	0	8.43
9	Load	0.001322	7.30
10	Load	0.001518	5.74
11	Load	0.001944	5.35
12	Load	0.001059	4.37
13	Load	0.001837	6.06
14	Load	0.001571	3.98

9.2. Normal Operation Conditions

Optimal power flow calculations are performed in normal operation conditions (base case) using DIgSILENT software. The results are shown in Table 3, which includes buses voltage magnitude (|V|) and generated reactive power. In this case, the penetration levels of wind and photovoltaic power plants at normal conditions are 54% and 26%, respectively.

Table 3. Optimal Power Flow Result in Normal Operation Conditions.

Bus No.	Normal Operation		Contingency 1		Contingency 2	
	|V| PU	Qg MVAR	|V| PU	Qg MVAR	|V| PU	Qg MVAR
1	1.05	39.83	1.05	44.2	1.05	43.82
2	1.02	−5.21	1.02	−5.20	1.01	−5.15
3	0.99	15.50	0.99	15.50	0.97	15.50
4	1.00	0.00	1.00	0.00	0.98	0.00
5	1.00	0.00	0.99	0.00	0.98	0.00
6	1.00	−9.91	1.00	−9.60	1.00	−4.40
7	1.02	0.00	1.02	0.00	1.01	0.00
8	1.05	19.58	1.05	21.55	1.05	20.62
9	1.02	0.00	1.02	0.00	1.01	0.00
10	1.01	0.00	1.01	0.00	1.00	0.00
11	1.00	0.00	1.00	0.00	1.00	0.00
12	0.99	0.00	0.99	0.00	0.99	0.00
13	0.98	0.00	0.98	0.00	0.98	0.00
14	0.98	0.00	0.98	0.00	0.98	0.00

9.3. Contingency 1

The wind farm at bus 1 consists of two blocks, the first one has 142.5 MW installed capacity and the second one has 47.5 MW installed capacity. This contingency simulates the outage of the second block. The second block was producing 30 MW before outage.

The optimal load flow calculations were conducted at this contingency to achieve minimum power losses and the results are shown in Table 3. To achieve the results of the tertiary voltage control, a coordinated secondary voltage control was applied in order to make the rest of wind farm and PV park support the pilot bus 5 to reach its optimal value. The coordinated PID controller parameters designed by GA using MATLAB/SIMULINK at this condition were found to be $K_P = 0.74$, $K_I = 0.23$, and $K_D = 0.00184$. The GA takes 14 iterations to converge. Figure 10 shows that SecVC has directed bus 5 voltage to its optimal steady-state value given in Table 3 which is 0.994 p.u. Figures 11 and 12 show the generated reactive power response from the wind farm and PV park, respectively, to control the load voltage.

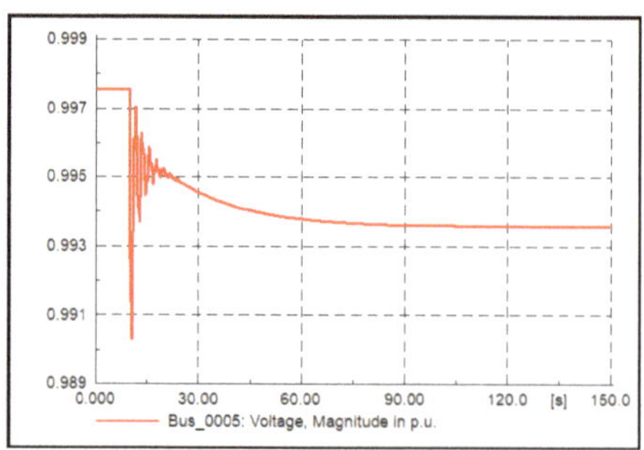

Figure 10. Bus 5 voltage in contingency 1.

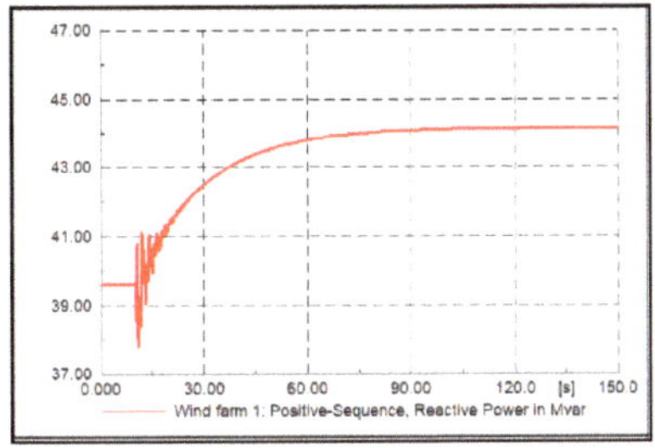

Figure 11. Reactive power of the first block of wind farm at bus 1 in contingency 1.

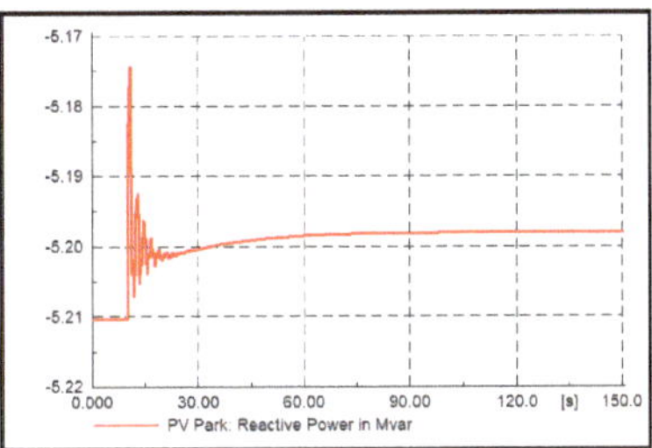

Figure 12. PV Park at bus-2 reactive power in contingency 1.

Figures 13 and 14 show the deviation in frequency and wind farm active power. PV plants did not contribute in the real power sharing process, because the frequency is in the dead band region.

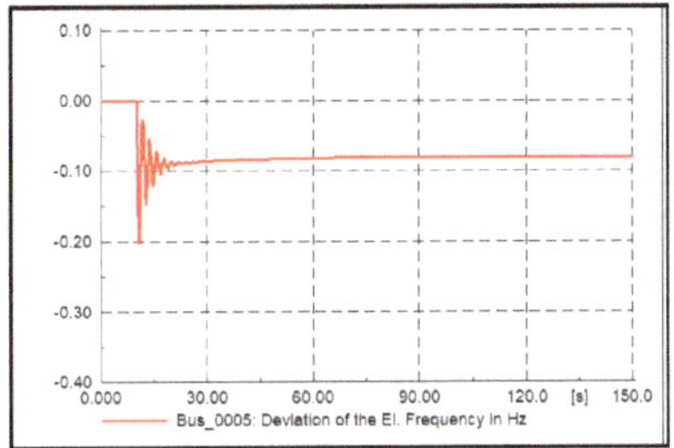

Figure 13. System frequency deviation in contingency 1.

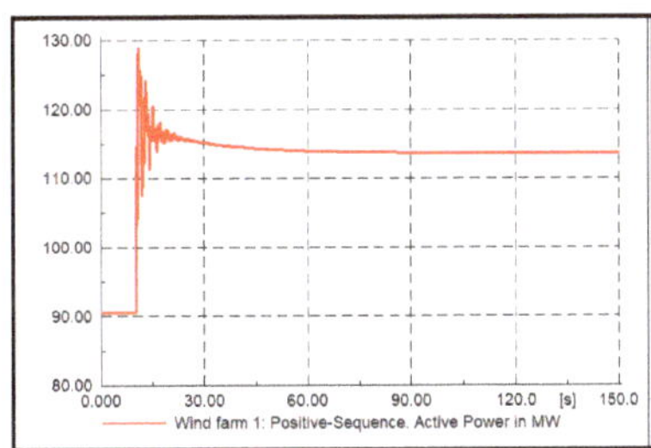

Figure 14. Active power of the first block of wind farm at bus 1 in contingency 1.

9.4. Contingency 2

Now, a short-circuit for 120 ms occurs on line 1–2 followed by line outage, in addition the load at bus 5 increases by 25% after 110 s. Optimal power flow calculations were performed at these events to achieve minimum power losses and the results are listed in Table 3. The secondary PID controller parameters are found to be $K_P = 1.7$, $K_I = 0.93$, and $K_D = 0.075$, and the control signal is directed to the wind farm installed at bus 1 to apply reactive power control to achieve the optimal voltage value at the pilot bus. Figure 15 shows the pilot bus voltage response, while Figure 16 shows the reactive power generated from the wind farm. Figure 17 shows system frequency deviation and Figure 18 shows the active power response of the wind farm; noting that the real power of PV park will be constant as the frequency deviation is within the dead band. The reactive current of the PV park during the short circuit remains constant, because the voltage of PCC did not reach below 0.9 pu in this case.

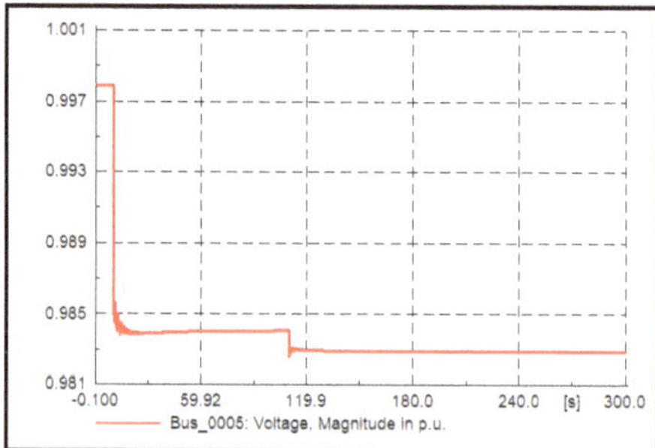

Figure 15. Bus 5 Voltage in contingency 2.

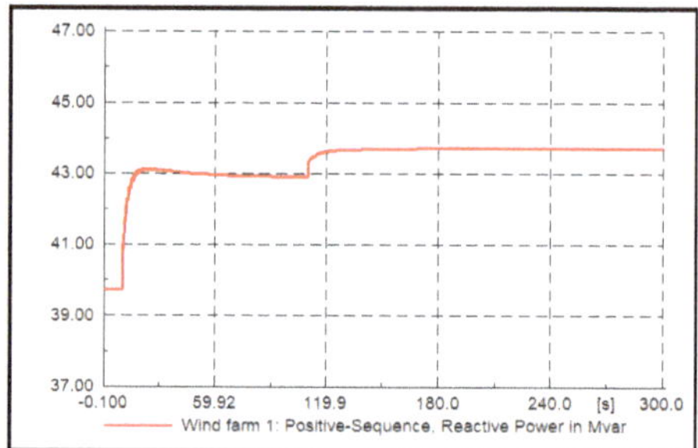

Figure 16. Wind farm reactive power in contingency 2.

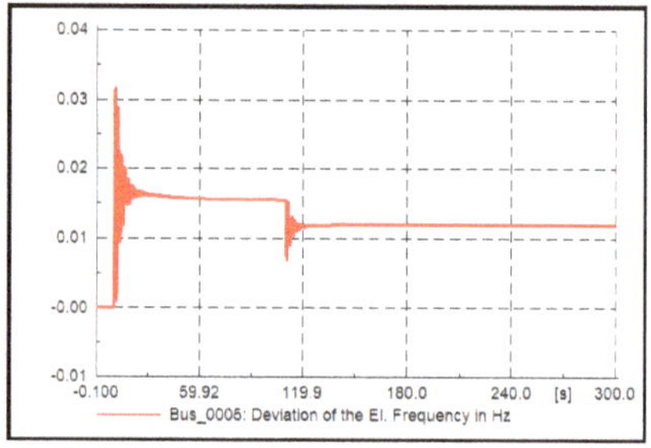

Figure 17. System frequency deviation in contingency 2.

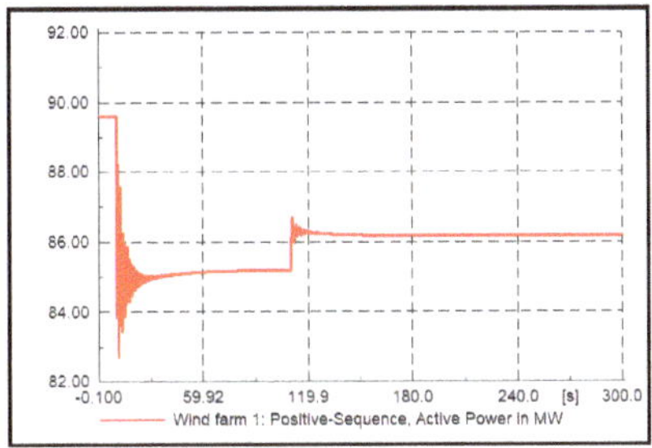

Figure 18. Wind farm active power in contingency 2.

9.5. Comparison of Power Losses and Voltage Index

Table 4 shows a comparison of system performances using proposed SecVC, droop SecVC and without SecVC in terms of total power losses and voltage deviation index (VDI) as illustrated in (16) [30]. The results show that the system proposed SecVC has better performance in terms of lower active losses and lower voltage deviation index through controlling the reactive power.

$$VDI = \sqrt{\frac{1}{n_L}\sum_{i=1}^{n_L}(V_{in} - V_i)^2} \tag{16}$$

where n_L is the number of load buses in the power system, which is equal to nine buses in the studied 100% renewable system, V_{in} is the nominal voltage of the load bus, and V_i is the actual load bus voltage.

Table 4. Comparison Between System Performance Using Proposed, Droop and Without SecVC.

Contingency No.	Losses in MW			Voltage Index		
	Proposed SecVC	Droop SecVC	Without SecVC	Proposed SecVC	Droop SecVC	Without SecVC
1	8.4	10.1	11.2	0.024	0.051	0.087
2	7.9	8.3	9.0	0.019	0.037	0.066

10. Conclusions

The paper has investigated the application of coordinated secondary voltage control to power system contingencies in a smart grid with 100% renewable generations in order to avoid voltage violations. Controls of active and reactive powers are considered based on grid codes of countries with high penetration level of renewable energy sources. The results have shown that the wind and solar renewable systems can support voltage by controlling the reactive power through inverters and the dc links. The proposed technique also led the system to reach better voltage profile and less total power losses than droop control. The paper also proved that the wind farm can contribute to primary frequency control using droop control. The coordinated PID controller optimal parameters change at each operating condition due to the change of the network configuration. Neural network is used to select the optimal parameters of the controller at each case. The PMUs are essential in the proposed work in order to provide real time voltage measurements to be used as inputs to the neural network. The paper has also investigated the pilot bus selection by two different methods; namely, sensitivity and short-circuit analyses.

Author Contributions: Conceptualization, H.H.F. and O.H.A.; methodology, H.H.F., A.M.A.G. and O.H.A.; software, H.H.F.; validation, H.H.F. and O.H.A.; formal analysis, H.H.F. and O.H.A.; investigation, H.H.F. and O.H.A.; resources, H.H.F. and O.H.A.; data curation, H.H.F. and O.H.A.; writing—original draft preparation, H.H.F. and O.H.A.; writing—review and editing, H.H.F. and O.H.A.; visualization, H.H.F., A.M.A.G. and O.H.A.; supervision, A.M.A.G. and O.H.A. All authors have read and agreed to the published version of the manuscript.

Funding: This research received no external funding and the APC was waived.

Conflicts of Interest: The authors declare no conflict of interest.

Nomenclature

B	Charge in P.U.
f_i	Output frequency
f_o	Desired frequency
K_v	Voltage regulator integrator gain
K_{IQ}	Reactive power integrator gain
K_P	Proportional gain constant
K_I	Integral gain constant
K_D	Derivative gain constant
n_{p_i}	Frequency droop gain
P_G	Generated active power
P_{Gmax}	Maximum generated power
P_{Gmin}	Minimum generated power
P_D	Demand active power
Pline	Line power loading
$Pline_{max}$	Maximum line power loading
Q_G	Generated reactive power
Q_{Gmax}	Maximum reactive power limit
Q_{Gmin}	Minimum reactive power limit
Q_D	Demand reactive power
R	Resistance in P.U.
S	Apparent power
X	Reactance in P.U.
Vb	Bus voltage in P.U.
V_P	Pilot bus actual voltage
V_{pref}	Pilot bus reference voltage
V_{Max}	Maximum valve position
V_{Min}	Minimum valve position
i_d, i_q	d and q axis currents
q	Secondary voltage control signal (action).
v_d, v_q	d and q axis voltages
ΔV_P	Error in pilot bus voltage

Abbreviations

AVR	Automatic voltage regulator
DFIG	Doubly fed induction wind generator
GA	Genetic algorithm
PID	Proportional integral derivative controller
PriVC	Primary voltage control
PMU	Phasor measurement unit
PV	Photovoltaic
PWM	Pulse width modulation
SCADA	Supervisory control and data acquisition
SecVC	Secondary voltage control
TerVC	Tertiary voltage control
VAR	Volt-ampere reactive (unit of reactive power).
WAMS	Wide area measurement system
WF	Wind farm

Appendix A

Loads data of the applied 100% renewable energy system are listed in Table A1 and lines data are listed in Table A2.

Table A1. 100% Renewable 14 Bus System Load Data.

Bus	P_D	Q_D	Bus	P_D	Q_D
1	0	0	8	0	0
2	21.7	12.7	9	29.5	16.6
3	94.2	19	10	9	5.8
4	47.8	−3.9	11	3.5	1.8
5	7.6	1.6	12	6.1	1.6
6	11.2	7.5	13	13.5	5.8
7	0	0	14	14.9	5

Table A2. 100% Renewable 14 Bus System Line Data.

Line	From Bus	To Bus	R	X	B
1	1	2	0.01938	0.05917	0.0264
2	2	3	0.04699	0.19797	0.0219
3	2	4	0.05811	0.17632	0.0187
4	1	5	0.05403	0.22304	0.0246
5	2	5	0.05695	0.17388	0.017
6	3	4	0.06701	0.17103	0.0173
7	4	5	0.01335	0.04211	0.0064
8	5	6	0	0.25202	0
9	4	7	0	0.20912	0
10	7	8	0	0.17615	0
11	4	9	0	0.55618	0
12	7	9	0	0.11001	0
13	9	10	0.03181	0.0845	0
14	6	11	0.09498	0.1989	0
15	6	12	0.12291	0.25581	0
16	6	13	0.06615	0.13027	0
17	9	14	0.12711	0.27038	0
18	10	11	0.08205	0.19207	0
19	12	13	0.22092	0.19988	0
20	13	14	0.17093	0.34802	0

References

1. Bose, B.K. Power Electronics, Smart Grid, and Renewable Energy Systems. *Proc. IEEE* **2017**, *105*, 2011–2018. [CrossRef]
2. Bose, B.K. Artificial Intelligence Techniques in Smart Grid and Renewable Energy Systems—Some Example Applications. *Proc. IEEE* **2017**, *105*, 2262–2273. [CrossRef]
3. Bhalshankar, S.S.; Thorat, C.S. Integration of smart grid with renewable energy for energy demand management: Puducherry case study. In Proceedings of the 2016 International Conference on Signal Processing, Communication, Power and Embedded System (SCOPES), Paralakhemundi, India, 3–5 October 2016; pp. 1–5.
4. Alam, M.J.E.; Muttaqi, K.M.; Sutanto, D. A Multi-Mode Control Strategy for VAr Support by Solar PV Inverters in Distribution Networks. *IEEE Trans. Power Syst.* **2014**, *30*, 1316–1326. [CrossRef]
5. Golsorkhi, M.S.; Shafiee, Q.; Lu, D.D.C.; Guerrero, J.M.; Esfahani, M.S.G. A Distributed Control Framework for Integrated Photovoltaic-Battery-Based Islanded Microgrids. *IEEE Trans. Smart Grid* **2016**, *8*, 2837–2848. [CrossRef]
6. Hu, B.; Canizares, C.A.; Liu, M. Secondary and Tertiary Voltage Regulation based on optimal power flows. In Proceedings of the 2010 IREP Symposium Bulk Power System Dynamics and Control—VIII (IREP), Rio de Janeiro, Brazil, 1–6 August 2016.
7. Alvarez, S.R.; Mazo, E.H.L.; Oviedo, J.E. Evaluation of power system partitioning methods for secondary voltage regulation application. In Proceedings of the 2017 IEEE 3rd Colombian Conference on Automatic Control (CCAC), Cartagena, Colombia, 18–20 October 2017; pp. 1–6.

8. Corsi, S. Closure of "The Coordinated Automatic Voltage Control of the Italian Transmission Grid—Part I: Reasons of the Choice and Overview of the Consolidated Hierarchical System". *IEEE Trans. Power Syst.* **2006**, *21*, 445–446. [CrossRef]

9. Martins, N. The new cigre task force on coordinated voltage control in transmission networks. In Proceedings of the 2000 Power Engineering Society Summer Meeting (Cat. No.00CH37134), Seattle, WA, USA, 16–20 July 2002. [CrossRef]

10. Su, H.-Y.; Liu, C.-W. An Adaptive PMU-Based Secondary Voltage Control Scheme. *IEEE Trans. Smart Grid* **2013**, *4*, 1514–1522. [CrossRef]

11. Abdalla, O.H.; Ghany, A.A.; Fayek, H.H. Coordinated PID secondary voltage control of a power system based on genetic algorithm. In Proceedings of the 2016 Eighteenth International Middle East Power Systems Conference (MEPCON), Cairo, Egypt, 27–29 December 2016; pp. 214–219.

12. Bose, A. Smart Transmission Grid Applications and Their Supporting Infrastructure. *IEEE Trans. Smart Grid* **2010**, *1*, 11–19. [CrossRef]

13. Li, J.; Liu, F.; Li, Z.; Mei, S.; He, G. Impacts and benefits of UPFC to wind power integration in unit commitment. *Renew. Energy* **2018**, *116*, 570–583. [CrossRef]

14. Kumar, M.; Nallagownden, P.; Elamvazuthi, I. Optimal Placement and Sizing of Renewable Distributed Generations and Capacitor Banks into Radial Distribution Systems. *Energies* **2017**, *10*, 811. [CrossRef]

15. Romero, M.E.; Seron, M. Ultimate Boundedness of Voltage Droop Control with Distributed Secondary Control Loops. *IEEE Trans. Smart Grid* **2018**, *10*, 4107–4115. [CrossRef]

16. Lai, J.; Lu, X.; Monti, A. Distributed secondary voltage control for autonomous microgrids under additive measurement noises and time delays. *IET Gener. Transm. Distrib.* **2019**, *13*, 2976–2985. [CrossRef]

17. Gas and Power UK. Available online: https://gulfgasandpower.uk/blog/top-renewable-energy-generating-countries-in-the-world (accessed on 20 June 2020).

18. Kroposki, B.; Johnson, B.; Zhang, Y.; Gevorgian, V.; Denholm, P.; Hodge, B.-M.; Hannegan, B. Achieving a 100% Renewable Grid: Operating Electric Power Systems with Extremely High Levels of Variable Renewable Energy. *IEEE Power Energy Mag.* **2017**, *15*, 61–73. [CrossRef]

19. Blaabjerg, F.; Yang, Y.; Yang, D.; Wang, X. Distributed Power-Generation Systems and Protection. *Proc. IEEE* **2017**, *105*, 1311–1331. [CrossRef]

20. Denholm, P.; Margolis, R. *Energy Storage Requirements for Achieving 50% Solar Photovoltaic Energy Penetration in California*; National Renewable Energy Lab.: Golden, CO, USA, 2016.

21. REN21. Functional requirements in the power system (translated title, in Norwegian: Funksjonskravi kraftsystemet (FIKS)). In *Renewable Energy Policy Network for the 21st Century Annual Report 2017*; REN21: Paris, France, 2017.

22. Standardization Administration of the People's Republic of China. *Technical Requirements for Connecting Photovoltaic Power Station to Power Systems*; GB/T 19964-2012; Standardization Administration of the People's Republic of China: Beijing, China, 2012.

23. Zheng, Q.; Li, J.; Ai, X.; Wen, J.; Fang, J. Overview of grid codes for photovoltaic integration. In Proceedings of the 2017 IEEE Conference on Energy Internet and Energy System Integration (EI2), Beijing, China, 26–28 November 2017; pp. 1–6.

24. Abdalla, O.H. Technical requirements for connecting medium and large solar power plants to electricity networks in Egypt. *J. Egypt. Soc. Eng.* **2018**, *57*, 25–36.

25. Lou, G.; Gu, W.; Xu, Y.; Cheng, M.; Liu, W. Distributed MPC-Based Secondary Voltage Control Scheme for Autonomous Droop-Controlled Microgrids. *IEEE Trans. Sustain. Energy* **2017**, *8*, 792–804. [CrossRef]

26. Mahari, A.; Seyedi, H. Optimal PMU placement for power system observability using BICA, considering measurement redundancy. *Electr. Power Syst. Res.* **2013**, *103*, 78–85. [CrossRef]

27. Shrivastava, D.R.; Siddiqui, S.; Verma, K. Optimal PMU placement for coordinated observability of power system under contingencies. In Proceedings of the 2017 IEEE International Conference on Circuits and Systems (ICCS), Thiruvananthapuram, India, 20–21 December 2017; pp. 334–339.

28. Chakrabarti, S.; Kyriakides, E.; Eliades, D.G. Placement of Synchronized Measurements for Power System Observability. *IEEE Trans. Power Deliv.* **2008**, *24*, 12–19. [CrossRef]

29. Dracopoulos, D.C. Genetic Algorithms and Genetic Programming for Control. In *Evolutionary Algorithms in Engineering Applications*; Dasgupta, D., Michalewicz, Z., Eds.; Springer Science and Business Media LLC: Berlin, Germany, 1997; pp. 329–343.

30. Abdalla, O.H.; Al-Badwawi, R.; Al-Hadi, H.S.; Al-Riyami, H.A.; Al-Nadabi, A. Steady-State and Dynamic Performance of Oman Transmission System with Diesel-Engine Driven Distributed Generation. In Proceedings of the 46th International Universities Power Engineering Conference (UPEC 2011), Soest, Germany, 5–8 September 2011.

Article

Tidal Supplementary Control Schemes-Based Load Frequency Regulation of a Fully Sustainable Marine Microgrid

Hady H. Fayek [1,*] and Behnam Mohammadi-Ivatloo [2,3]

[1] Electromechanics Engineering Dept., Faculty of Engineering, Heliopolis University, Cairo 11785, Egypt
[2] Faculty of Electrical and Computer Engineering, University of Tabriz, Tabriz 5166616471, Iran; bmohammadi@tabrizu.ac.ir
[3] Department of Energy Technology, Aalborg University, 9200 Aalborg, Denmark
* Correspondence: hadyhabib@hotmail.com; Tel.: +20-1005472291

Received: 20 August 2020; Accepted: 24 October 2020; Published: 7 November 2020

Abstract: The world is targeting fully renewable power generation by the middle of the century. Distributed generation is the way to increase the penetration level of renewable energies. This paper presents load frequency control of a hybrid tidal, wind, and wave microgrid to feed an isolated island. This research is a step towards 100% renewable energy communities in remote seas/oceans islands. The wave and tidal generation systems model are presented. The study presents load frequency control through three supplementary control strategies: conventional integrators, fractional order integrator, and non-linear fractional order integrator. All the controllers of the microgrid are designed by using a novel black widow optimization technique. The applied technique is compared to other existing state-of-the-art algorithms. The results show that the black widow non-linear fractional integrator has a better performance over other strategies. Coordination between the unloaded tidal system and blade pitch control of both wind and tidal systems are adopted in the microgrid to utilize the available reserve power for the frequency support. Simulation and optimization studies are performed using the MATLAB/SIMULINK 2017a software application.

Keywords: marine microgrid; tidal generation system; black widow optimization; supplementary control; fractional integrator; non-linear fractional integrator; 100% renewable power generation

---

1. Introduction

The world is targeting 100% renewable power generation by the middle of this century [1]. Distributed renewable energy generation worldwide is increasing due to its low carbon dioxide emission and cost. The integration of renewable energies to the grid or high penetration level of renewables standalone systems suffers in terms of stability, power quality, and reliability [2]. Load frequency control plays an important role in terms of improving the power quality of the power systems, which operates with high variability in both loads and generation due to the presence of renewables [3].

Tidal energy is considered one of the promising renewable energy technologies in the 100% renewable energies dream. In [4], the researchers presented a feasibility study of tidal energy applications. Tidal power plants were always associated with offshore wind turbines to cover consumption needs. The integration of wind and tidal power plants to a grid or a power system were welcomed despite the frequency stability problems. With the increase of the penetration level of renewables in standalone systems and interconnected power systems, renewable energies must participate in the load frequency control process. The contribution is made in tidal and wind generators by deloading. Deloading can be defined as the operation of the wind or tidal system at a power below the maximum power to create a reserve. The reserve can be utilized to stabilize the power system for frequency regulation during active power mismatch between generation and demand [5]. However, using deloading only cannot drive to the observed reduction in frequency deviation, but involving inertia and damping controls can reduce the reduction [6].

There are a lot of technologies to convert tidal energy into electrical energy. The best known technologies are barrages, turbines, and fences. In this paper, turbines technology is assumed to be used to convert tidal energy into a mechanical one. Through doubly-fed induction generators, the mechanical energy is converted to an electric one. In [7], the researchers presented a complete study on the dynamic behavior of pitch and stall regulated tidal turbines. In [8], the effect of integration tidal power plant on a real grid is discussed. Maximum power point tracking under different tides speed is presented in [9]. In [10,11], hybrid offshore wind tidal generating system is presented. Based on the literature review, there is not much focus on load frequency control by using the deloading of tidal generators. In this work, the deloading is applied by using inertia, damping, and various supplementary control schemes.

Wave energy has good potential in the 100% renewable power generation goal. It can satisfy more than 10% of the total global demand for electricity [12]. The structure of the Archimedes wave swing is presented in [13]. In this paper, wave energy was simulated as an uncontrollable generation system.

Control of microgrids is an important topic addressed by many researchers and applied using different optimization techniques. In [14], a particle swarm algorithm (PSO) is presented. In [15], a genetic algorithm (GA) is presented. In [16], the teaching and learning algorithm (TLBO) is presented, and in [17], harmony search is presented. The black widow optimization algorithm is presented for the first time in [18] by Hayyolalam and Kazem as a novel meta-heuristic approach for solving engineering optimization problems.

Load frequency control was previously applied by using different control schemes: proportional integral (PI) control [19], proportional integral derivative (PID) control [20], H-infinity [21], fractional order proportional integral derivative (FOPID) control [22], and non-linear proportional integral derivative (NPID) control [23]. The researchers proved the simplicity of PID, while better performance can be achieved through FOPID and NPID. In this paper, the supplementary control is designed to be a non-linear fractional integrator. At the same time, other controllers will be set to be PID to ensure both simplicity and quick response to a frequency deviation.

In [24], the research presents load frequency control of standalone tidal and diesel microgrid. The research presents a contribution through supplementary control from tidal generation to demand change.

In [25], the research load frequency control for wind-diesel generation microgrid using the D-partition method (DPM). The research provided a single step /simplistic computing method for calculating the PI controller parameters of a dynamic system, such as the microgrid system comprising of the renewable energy sources without any further requirements of retuning.

In [26], a novel fractional-order model predictive control technique is presented to track the optimal frequency of a standalone microgrid through including fractional-order integral cost function into model predictive control (MPC) algorithm.

In [27], In this research, a simulation and control of tidal generation system has been presented. A fuzzy system has been used for the pitch controller to properly modify the gains of the PID at different tidal inputs.

The main contributions of this paper are:

1. Simulation and control of a 100% renewable energy microgrid including tidal, wave and offshore wind hybrid generation system.
2. The effect of different supplementary control schemes in terms of the integrator, fractional integrator, and non-linear fractional integrator on the dynamic performance of load frequency control (LFC) is examined.

3. Design of hybrid system controllers and tidal supplementary controller by using a novel black widow optimization technique and comparing it with other state-of-art optimization techniques.

The paper is organized as follows. Section 2 presents the modelling of the microgrid, Section 3 illustrates the applied control schemes. Section 4 presents the controller's design process, Section 5 presents simulation results, and Section 6 is the main conclusion of this research.

2. Microgrid Modelling

The world is targeting 100% renewable energies by 2050. This paper presents a study on a standalone microgrid operated by fully sustainable marine generating systems, which are tidal, wind, and wave generators, as illustrated in Figure 1. The system's mathematical descriptions are presented in the following subsections.

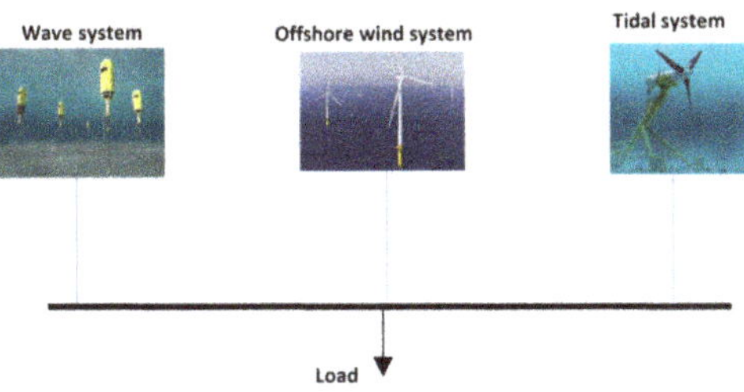

Figure 1. Marine microgrid.

2.1. Modeling of the Tidal Generating System

Both wind generating systems and tidal generating systems have similar operation and control principles. The difference between the two systems is that the tidal speed and nominal turbine size are less than for the wind. The wind rated speed ranges from 12–15 m/s and the tidal one from 2–3 m/s. The mechanical power output (P_T) can be modeled as illustrated in (1) [4].

$$P_T = \frac{1}{2}\mathcal{P}AV^3C_P(\gamma,\beta) \tag{1}$$

where $\mathcal{P}$, A, and V are the seawater density, turbine blades area, and tidal speed flow, respectively; C_P is the power coefficient in terms of tip speed ratio (γ) and blade pitch control angle (β); while d_1, d_2, d_3, d_4, d_5, d_6, and d_7 are $C_P(\gamma,\beta)$ equation parameters.

$$C_P(\gamma,\beta) = \left(\frac{d_1d_2}{d_6\beta+\gamma} - \frac{d_1d_7}{\beta^3+1} - d_1d_3\beta - d_1d_4\right)\exp\left(\frac{-d_5}{d_6\beta+\gamma} + \frac{d_5d_7}{\beta^3+1}\right) \tag{2}$$

$$\gamma = \frac{\omega R}{V} \tag{3}$$

Such that ω and R are the rotational speed of the tidal blades and radius of the blades, respectively. The tidal generating system has four modes of operation, as illustrated in Figure 2 and Table 1.

Turbine power variations are governed by the pitch control system comprising of a PID controller, which gets its input as error between measured turbine rotor speed and change in reference speed. The power varies with blade pitch control to what achieves the relationship between power output and rotor speed, as illustrated in Figure 2 [24].

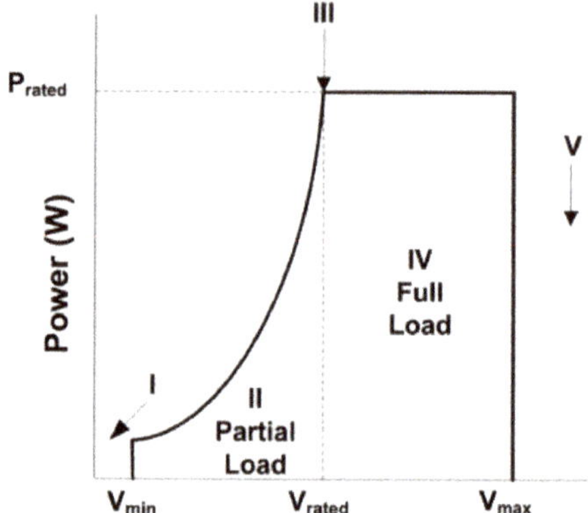

Figure 2. Tidal turbine modes of operation [24].

Table 1. Tidal turbine modes of operation [21].

Mode No.	Condition	Operation
I	$V \leq V_{min}$	No power generation with pitch angle setting 90 degrees.
II	$V_{min} < V \leq V_{rated}$	Optimum power extraction from the turbine to reach optimum efficiency, Blade pitch angle is set at 4 degrees in this work
III	$V_{rated} < V \leq V_{max}$	Constant power operation turbine, blade pitch angle is varied from 4 degrees to 90 degrees to avoid overload.
IV	$V > V_{max}$	No power output and blade pitch angle is set at 90 degrees.

Participation for frequency support is a must for 100% renewable generating systems. To ensure that tidal generating systems participate in frequency control, it is necessary to operate them at a level well below the maximum power point. This phenomenon is called deloading such that the power output of the system is varied between deloading power P_{del} and maximum power P_{max} as shown in Figure 3. This could happen by varying the rotor speed from deloading speed (ω_{del}) to nominal or rated speed (ω_r). The maximum deloading percentage x can be calculated based on the maximum allowed rotor speed for the generator, as illustrated in Figure 4.

$$P_{del} = (1 - x)P_{max} \tag{4}$$

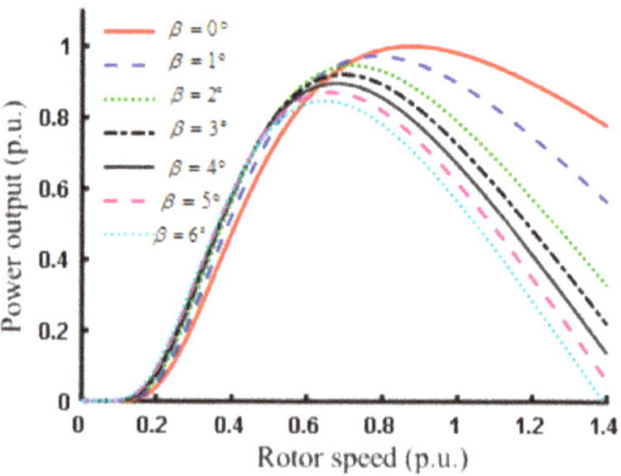

Figure 3. Tidal rotor speed vs. power output at different blade pitch angles [24].

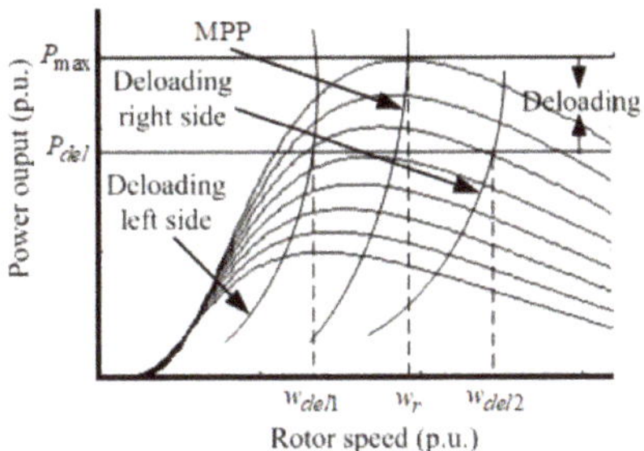

Figure 4. Deloading philosophy of tidal turbine [24].

Coordination between blade pitch control system and deloading system is required such that with the increase of the deloading percentage, the rotor speed may increase. At that instant, the blade pitch control system should adjust the blade pitch angle. The dynamic power reference (P_r) at a specific rotor speed and reference speed (ω_{ref}) can be defined as:

$$P_r = P_{del} + (P_{max} - P_{del})\left(\frac{\omega_{del} - \omega_m}{\omega_{del} - \omega_r}\right) \tag{5}$$

$$\omega_{ref} = \frac{P_r}{T_m} \tag{6}$$

where the measured mechanical toque is T_m.

To improve the frequency response, an additional signal ΔP_{id} will be added to the reference power output, as shown in Figure 5. The additional signal is coming from combined inertia and damping non- conventional machine equivalent controller. In other words, the frequency deviation and rate of change of frequency are represented in two signals, as illustrated in (7).

$$\Delta P_{id} = -D_1 \Delta f - M\frac{\partial \Delta f}{\partial t} \tag{7}$$

where D_1 and M are the additional damping and the additional inertia, respectively. To improve the response of the tidal system to the frequency deviation, different control schemes are compared to be added in parallel to D_1 and M.

To drive the speed to track the reference speed, a PID control is applied, as illustrated in (8) and shown in Figure 6.

$$\Delta P_\omega = K_{\omega TP}\Delta\omega_e + K_{\omega TI}\int \Delta\omega_e dt + K_{\omega TD}\frac{d\Delta\omega_e}{dt} \tag{8}$$

where $\Delta\omega_e$ is the error in speed deviation while $K_{\omega TP}$, $K_{\omega TI}$, and $K_{\omega TD}$ are the speed controller parameters. Thus, the change of the output power of the tidal generating system (ΔP_{Tidal}) can be formulated as (9).

$$\Delta P_{Tidal} = \frac{\partial P}{\partial \omega}\Delta\omega + \frac{\partial P}{\partial v}\Delta v + \frac{\partial P}{\partial \beta}\Delta\beta \tag{9}$$

where $\frac{\partial P}{\partial \omega}$, $\frac{\partial P}{\partial v}$, and $\frac{\partial P}{\partial \beta}$ are the change of tidal power with respect to the specific variation of turbine rotor speed, tidal speed, and blade pitch angle, respectively.

Tidal generating system parameters values are illustrated in Appendix A.

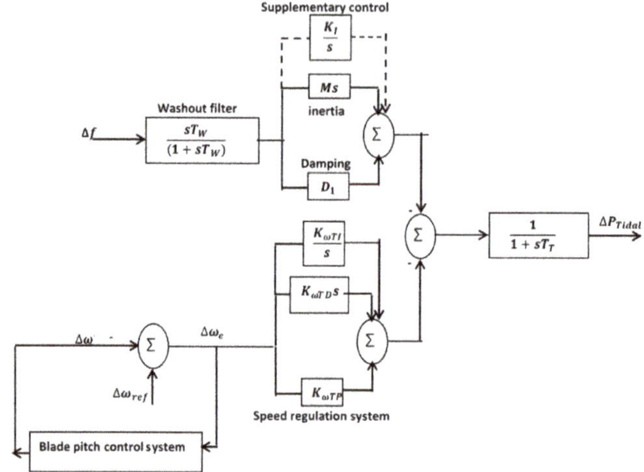

Figure 5. Tidal generation system model.

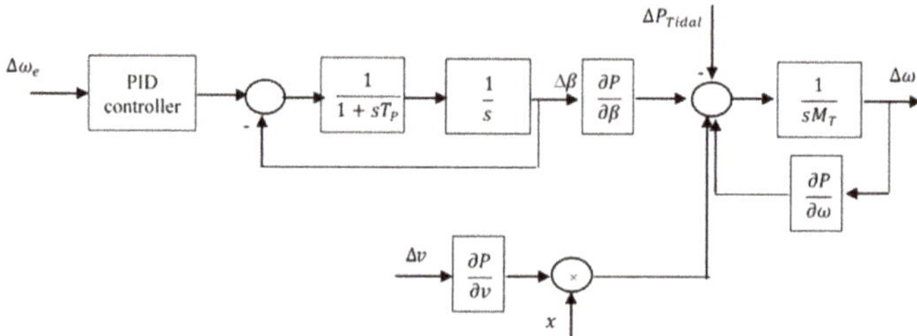

Figure 6. Blade pitch control of the tidal system.

2.2. Modeling of Wave Generating System

The wave generating system is one of the promising marine sustainable energy systems, which is not yet widely covered in the research. The wave system is assumed to be coupled with a permanent magnet synchronous generator. The following equations describe the dynamics of the system in terms of force (F_W) and velocity (V_W) [12,13].

$$V_W = \frac{dx}{dt} \tag{10}$$

$$F_W = m_{ft}\frac{dV_W}{dt} + (\beta_G + \beta_W)V_W + k_c x \tag{11}$$

where x is the floater and translator displacement, m_{ft} is the total mass, β_G is the damping constant of the generator, and β_W is the damping constant of the wave swing.

The system is simulated as a first-order generator, first-order converter, and first-order inverter, as illustrated in (12).

$$G_{wave} = \frac{K_{wave}}{1 + sT_{wave}}\frac{1}{1 + sT_{conv}}\frac{1}{1 + sT_{inv}} \tag{12}$$

where G_{wave} is the transfer function of the system, and K_{wave} and T_{wave} are the gain and time constants of the wave generator. T_{conv} and T_{inv} are the time constants of converter and inverter, respectively.

2.3. Modeling of Wind Generating System

The offshore wind generating system applied in this study is assumed to participate in frequency deviation by PID blade pitch control. The system is shown in Figure 7 and modeled in [22,25].

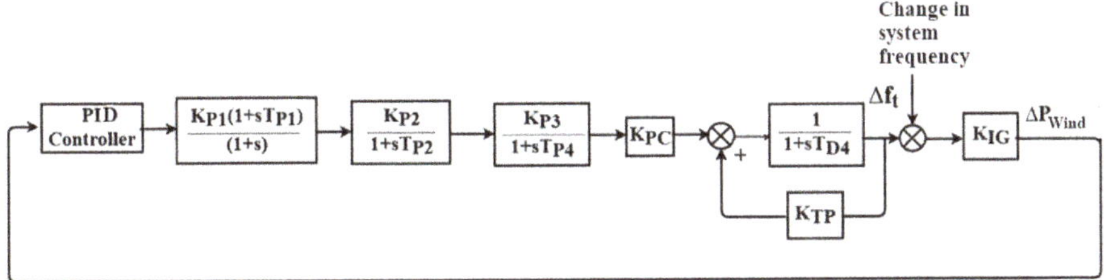

Figure 7. Offshore wind generating system.

2.4. Modeling of Microgrid

The microgrid is now formed from three marine systems. The change of the total generated power (ΔP_G) can be formulated as (13).

$$\Delta P_G = \Delta P_{Tidal} + \Delta P_{Wave} + \Delta P_{Wind} \tag{13}$$

The difference between the change in generation and the change in load is ΔP_{Gd}

$$\Delta P_{Gd} = \Delta P_G - \Delta P_D \tag{14}$$

The transfer function of the power system (G_{PS}) in terms of changes in system frequency (Δf) and ΔP_{Gd} is illustrated in (15). D represents the frequency dependency of the load while H is the microgrid moment of inertia.

$$G_{PS} = \frac{\Delta f}{\Delta P_{Gd}} = \frac{1}{2Hs + D} \tag{15}$$

3. Controllers

3.1. Blade Pitch Controllers

Wind and tidal turbines have blade pitch controllers, which are selected to be proportional integral derivative (PID) controllers. PID controllers are the most widely used tool to minimize the error. The input of the PID controller is the error, and the output of the controller is the control action received by the actuators. Appropriate power injection is usually done through PID control, which usually leads to a minimum frequency deviation:

$$G_{PID} = K_P + \frac{K_I}{s} + K_D s \tag{16}$$

where K_P is the proportional gain, K_I is the integral gain, and K_D is the derivative gain.

3.2. Tidal Speed Controller

The tidal speed controller is selected to be PID controller, like the blade pitch controller.

3.3. Tidal Supplementary Control Schemes

Different tidal supplementary control schemes are applied to minimize the frequency deviation of the microgrid, which are as follows:

(i)   Integrators (I scheme), which have the transfer function $G_I = \frac{K_I}{s}$

(ii)   Fractional integrators (FI scheme), which have the transfer function $G_{FI} = \frac{K_I}{s^\lambda}$

(iii)   Non-linear fractional integrators (NFI), which have the output $U_{NFI}(s) = \left(\frac{e^{(GxE)} + e^{-(GxE)}}{2} \frac{K_I}{s^\lambda} \right) E(s)$

where λ is the fractional-order operator of the integrator while G is the non-linearity gain, E is the error, and U_{NFI} is the output of the non-linear fractional integrator. Figure 8 shows the structure of NFI.

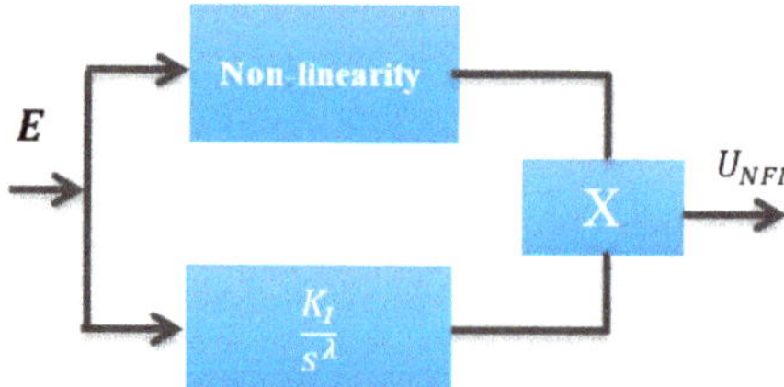

Figure 8. Non-linear fractional order integrator.

4. Control Design

4.1. Optimization Problem Definition

The design of wind and tidal turbines with blade pitch controllers, tidal speed controllers, and integrator based tidal supplementary control schemes will be performed for test operation condition using the following algorithms:

i.   Black widow
ii.   Quasi oppositional harmony search
iii.   Teaching and learning-based optimization
iv.   Particle swarm optimization
v.   Genetic algorithm

The best performance technique will be used for the rest of the subjected disturbances. The design is made in MATLAB to minimize frequency deviation. The optimization problem is described as follows:

- Objective Function (O): Minimization of integral absolute error (IAE) (O_1) and minimization of integral time absolute error (ITAE) (O_2):

$$O_1 = \min \int_0^T \Delta f \, dt \tag{17}$$

$$O_2 = \min \int_0^T t * \Delta f \, dt \tag{18}$$

$$O = 0.5 * O_1 + 0.5 * O_2 \tag{19}$$

- Variables: PID control parameters, tidal additional damping, and inertia, in addition to tidal supplementary control schemes parameters.
- Constraints: G and λ.

The multi-objective function is established such that weights of the ITAE and IAE are equal. The tidal supplementary control schemes parameters differ from one to another. In fractional integrators, there are two variables (K_I and λ) and the constraint considered in its design is $0 \leq \lambda \leq 1$

only. In non-linear fractional integrators, there are three variables (K_I, G, and λ) and the constraints considered are $0 \leq \lambda \leq 1$ and $0 \leq G \leq 1$ only.

4.2. Black Widow Optimization

Recently, due to the complexity of renewable energy-based power systems, the need for a viable meta-heuristic method has emerged. Nature-inspired optimization algorithms are used widely for solving power system problems in an easy and flexible way. Their method is inspired by the unique mating behavior of the black widow spider. The proposed optimization technique includes the cannibalism stage, which omits inappropriate fitness from the selection circle, so that convergence comes earlier. Figure 9 illustrates the proposed algorithm steps, which are:

1. Initial population: This is used in each optimization technique; it has other names, like chromosomes in the genetic algorithm and particle position in the particle swarm algorithm. In the black window, it has the name widow. To start the optimization, a candidate widow matrix with size $N_{pop} \times N_{var}$ is generated, where N_{var} represents the solution of the problem array while N_{pop} represents the number of populations.
2. Procreate: In this step, an array called α is created such that the offspring is produced through (20) and (21):

$$y_1 = \alpha \times x_1 + (1 - \alpha) \times x_2 \tag{20}$$

$$y_2 = \alpha \times x_2 + (1 - \alpha) \times x_1 \tag{21}$$

    where x_1 and x_2 are parents, while y_1 and y_2 are offspring. The process repeated every $\frac{N_{var}}{2}$.
3. Cannibalism: There are three kinds of cannibalism: (a) sexual cannibalism, where the female black widow eats her husband; in the algorithm, we can identify male and female through their fitness function; (b) sibling cannibalism, where the strong spider siblings eat their weaker siblings; in the algorithm, the cannibalism rating is set according to the determined number of survivors; and (c) baby cannibalism, where baby spiders eat their mother; in the algorithm, strong and weak spider siblings are recognized through fitness value.
4. Mutation: Random selection of Mutepop number of individuals. Mutepop is calculated according to the mutation rate.
5. Convergence: The same concept of many algorithms comes in the proposed algorithm; three stop conditions may be used: (a) a predefined number of iterations; (b) the fitness value is almost constant for several iterations; and (c) the desired accuracy is reached.

Three parameters must be set for obtaining the desired results: procreating rate (pp), cannibalism rate, (CR) and mutation rate (pm). In this research, the parameters are selected as pp = 0.62, CR = 0.46, and pm = 0.4.

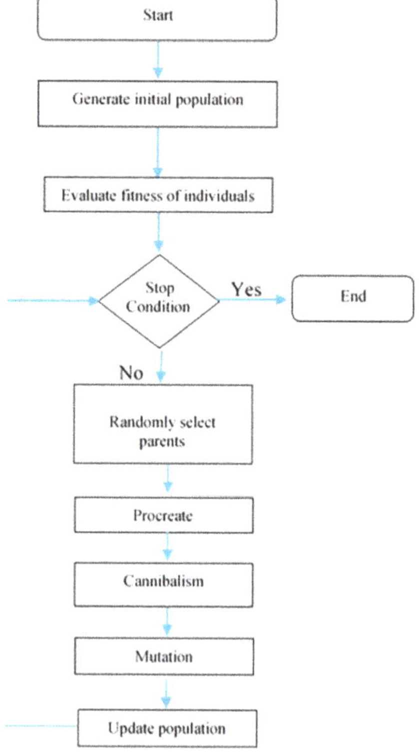

Figure 9. Black widow flow chart.

5. Simulation Results

In this part, the dynamic performance of the microgrid in terms of load frequency control is presented under different operating conditions. It is assumed that the tidal system is operating at tide speed 2.4 m/sec having a deloading effect with a 3o blade pitch control angle. The microgrid is simulated and optimization processes are applied on MATLAB/SIMULINK 2017a works in a Core i5, 2.50 GHz Samsung laptop with 6 GB of RAM. The parameters of all controllers in the wind and tidal systems are optimized using black widow optimization and its code is established using MATLAB 2017a. In order to establish the supremacy of black widow for the present work, system performance for 1% step load increase using integrator tidal supplementary control scheme is compared to other state-of-the-art optimization methods. Figures 10–12 show the frequency deviation, change in tidal power, and change in wind power. Table 2 shows a comparison between the performance of each optimization technique on ITAE, IAE, $\int(\Delta f)^2$, transient response of Δf in terms of undershoot (U_{sh}), overshoot (O_{sh}), settling time (t_s) in addition to peak time (t_p), and the number of iterations performed using each optimization technique. The results show that the black widow algorithm has the best performance over other algorithms. Therefore, the tests applied on the system to compare between different supplementary controllers will be applied using the black widow algorithm only. The dynamic study of the microgrid in terms of load frequency control, under the action of black widow tuned control schemes installed in the studied system, was subjected to the following tests:

Test 1: A step increase in the demand;
Test 2: Real demand variation at a certain day;
Test 3: Sinewave variation of the wave generation system;

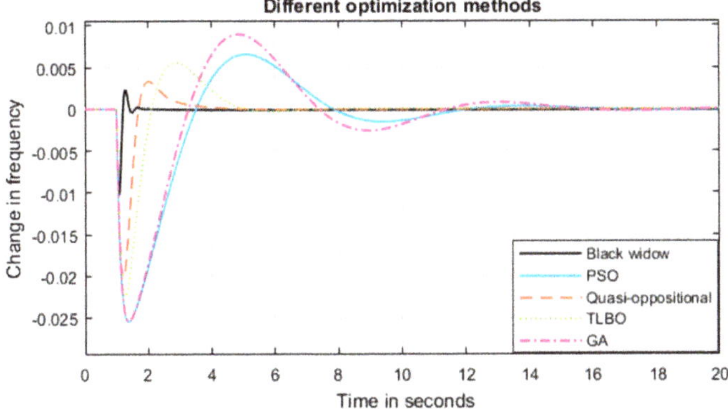

Figure 10. Change in frequency using different optimization techniques.

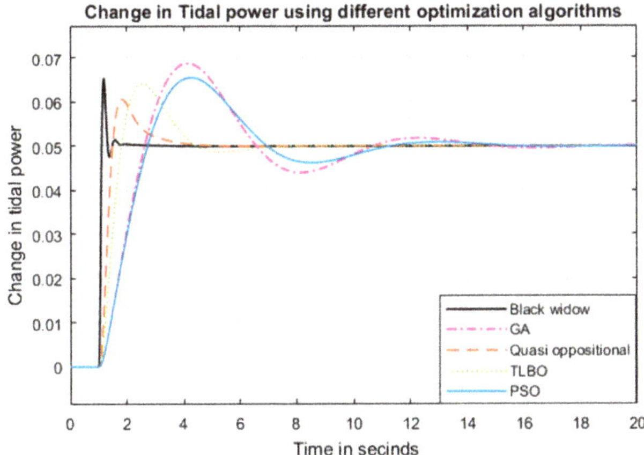

Figure 11. Change in tidal power using different optimization techniques.

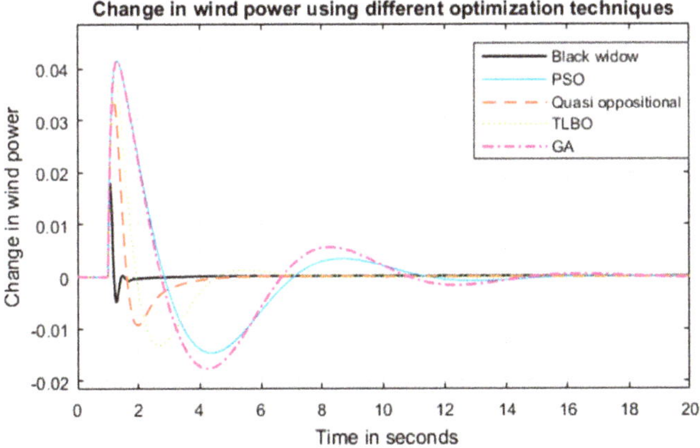

Figure 12. Change in wind power using different optimization techniques.

Table 2. Comparison of optimization methods.

Method	ITAE	IAE	$\int(\Delta f)^2 {*} 10^{-6}$	Number of Iterations	Transient Response of Δf			
					U_{sh}	O_{sh}	t_s	t_p
Black widow	0.0059	0.0011	0.39	15	−0.01	0.0025	1.75	1.5
Quasi-oppositional	0.0064	0.0016	0.42	21	−0.02	0.003	3.5	2
TLBO	0.0071	0.0019	0.48	18	−0.022	0.006	4.5	2.8
PSO	0.0085	0.0026	0.61	25	−0.025	0.007	15	4
GA	0.0089	0.0027	0.64	23	−0.025	0.009	15	4

5.1. System Performance under Test 1

In this test, the studied microgrid is subjected to a step increase in demand by 10% 1 second after starting the simulation. Four control strategies have been carried out on the system:

Strategy a: System without supplementary control
Strategy b: System with integrator supplementary control
Strategy c: System with fractional integrator supplementary control
Strategy d: System with non-linear fractional integrator supplementary control.

Figures 13–15 show the frequency deviation, change in tidal power, and change in wind power when the system is subjected to *Test 1*. The results show that the non-linear fractional integrator achieved a better performance than the fractional integrator, followed by the conventional integrator. To avoid mechanical oscillations, the optimal parameters of the supplementary control, added damping (D_1) and added inertia (M) in addition to the rest of the controllers of the system, are tuned at the same time to get the optimal parameters for the whole system considering the mechanical oscillations. If the integrator is not inserted in the system, there is a steady-state error and a very long settling time. System controllers and different supplementary control schemes parameters are illustrated in Table 3. Table 4 illustrates the performance of the microgrid at different frequency dependencies of the load (D); the results show that as D increases, the performance of the microgrids gets better in terms of ITAE, IAE, and settling time: in other words, as the loads used are more frequency dependent, the microgrid performance improves.

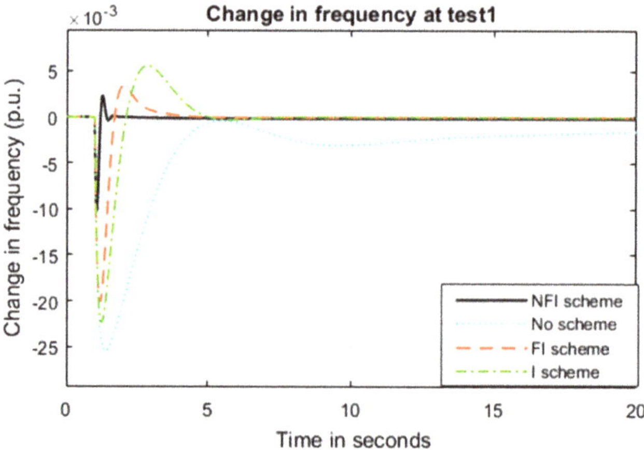

Figure 13. Change in frequency when the system was subjected to test 1.

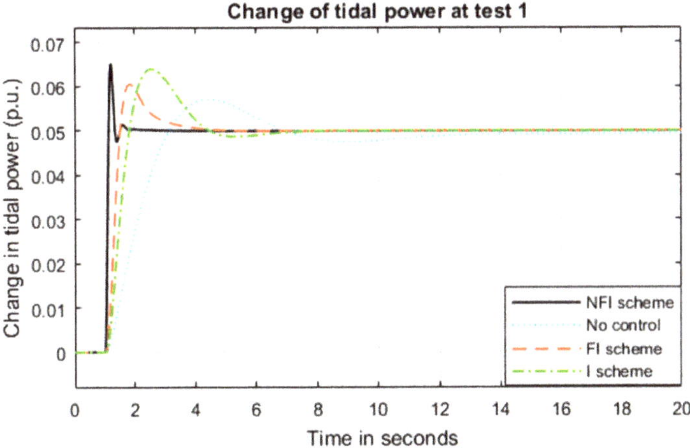

Figure 14. Change in tidal power when the system was subjected to test 1.

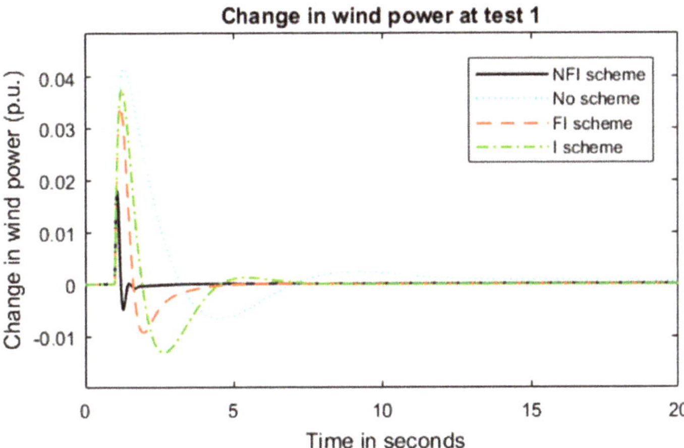

Figure 15. Change in wind power when the system was subjected to test 1.

Table 3. Microgrid control schemes parameters in test 1.

Control Scheme	M	D_1	Supplementary Control			Tidal Blade Pitch Controller			Tidal Speed Regulator			Wind Blade Pitch Controller		
			K_I	λ	G	K_P	K_I	K_D	$K_{\omega TP}$	$K_{\omega TI}$	$K_{\omega TD}$	K_P	K_I	K_D
No scheme	150	147	none	none	none	10	3	0.4	50	17	12	16	5	0.3
I scheme	122	85	10	none	none	17	14	11	14	5	4	17	1	0.2
FI scheme	146	75	7	0.43	none	11	9	0.16	6	4	1.4	8	5	2
NFI scheme	98	56	13	0.64	0.72	14	7	1.14	8	3	0.57	21	17	8

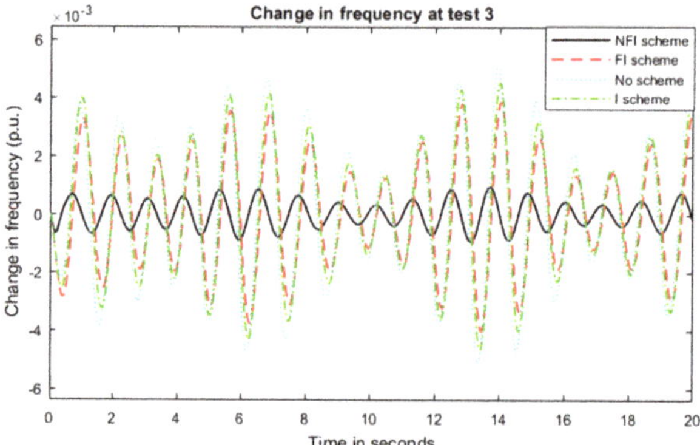

Figure 20. Change in frequency in test 3.

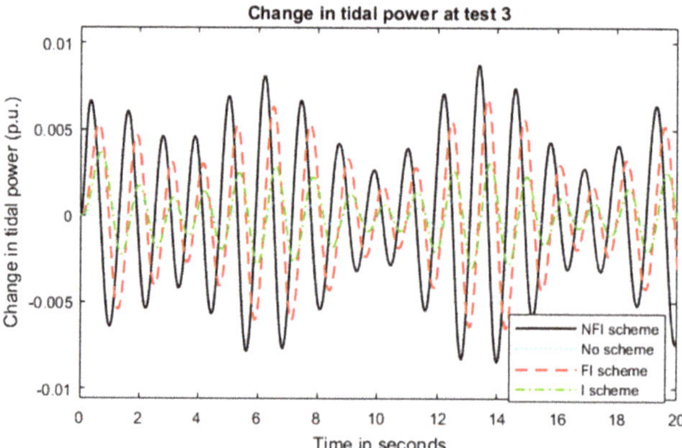

Figure 21. Change in tidal power in test 3.

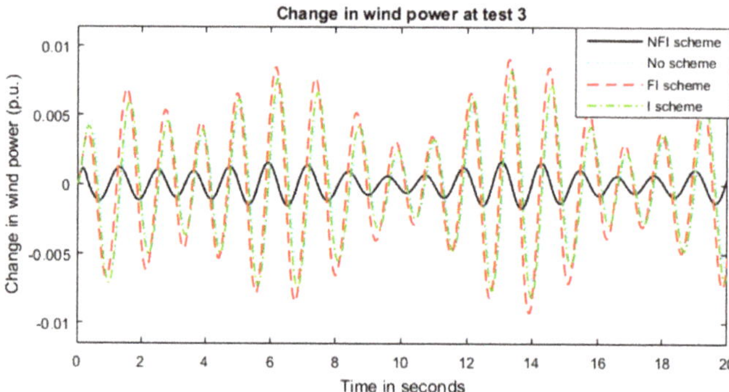

Figure 22. Change in wind power in test 3.

6. Conclusions

This paper has presented the load frequency control of a 100% renewable energy marine microgrid in terms of wind, tidal, and waves generators. The results show that using a tidal supplementary controller in the presence of an integrator drives the microgrid to zero steady-state frequency deviation in different operating conditions. The results show that the contribution of the tidal supplementary controller to the load or generation variation is more effective than the wind blade pitch controller in different operating conditions. The results also proved that the proposed non-linear fractional order integrator (NFI) based supplementary control achieves better performance than fractional order integrator (FI) and conventional integrator (I) schemes in different operating conditions. The results also proved that the FI control scheme drives the system to a better performance than the I control scheme. The results also proved that the controller design using a black widow optimization algorithm drives the system to a better performance than other existing state-of-the-art algorithms, in terms of ITAE, IAE, number of iterations, and change in frequency transient response. The results also show that using an NFI control scheme will lead the system to a better performance when subjected to sinusoidal wave power generation than will FI and I control schemes. The paper presents a technique to solve the frequency deviation in sea/ocean isolated microgrid. To practically apply this technique, integration between tidal, wind and wave technologies and software and microcontroller facilities, are required.

Author Contributions: Conceptualization, H.H.F.; methodology, H.H.F.; software, H.H.F.; validation, H.H.F.; formal analysis, H.H.F.; investigation, H.H.F.; resources, H.H.F. and B.M.-I.; data curation, H.H.F.; writing—original draft preparation, H.H.F. and B.M.-I.; writing—review and editing, H.H.F. and B.M.-I.; visualization, H.H.F. All authors have read and agreed to the published version of the manuscript.

Funding: This research received no external funding.

Conflicts of Interest: The authors declare no conflict of interest.

Appendix A

Table A1. Marine microgrid system parameters.

System	Parameters
Tidal	Capacity: 1 MW, rated rotor speed (ω_r) = 13 rpm, tidal speed (V) = 2.4 m/s, TSR = 6.1, rotor radius (r) = 11.5 m, rotor blades = 3, blade length = 10.6 m, rotor position = upstream, M_T = 0.3878 s, T_P = 0.01 s, T_T = 0.08 s, T_w = 6 s, angle limits: minimum = 0° and maximum = 90°, d_1 = 0.18, d_2 = 85, d_3 = 0.38, d_4 = 10.2, d_5 = 6.2, d_6 = 0.025, d_7 = −0.043
Wave	Capacity: 1 MW, K_{wave} = 1, T_{wave} = 0.3 s, T_{inv} = T_{conv} = 0.01 s
Offshore wind	Capacity: 1 MW, K_{p1} = 1.250, K_{p2} = 1.000, K_{p3} = 1.400, K_{TP} = 0.0033, K_{IG} = 0.9969, K_{PC} = 0.0800, T_{p1} = 0.6000 s, T_{p2} = 0.0410, T_{p3} = 1.000, T_W = 4.000.
Microgrid	H = 5, D = 0.8, f = 50 Hz

References

1. Abdalla, O.H.; Fayek, H.H.; Abdel Ghany, A.M. Secondary Voltage Control Application in a Smart Grid with 100% Renewables. *Inventions* **2020**, *5*, 37. [CrossRef]
2. Fayek, H.H. Voltage and Reactive Power Control of Smart Grid. Ph.D. Thesis, Helwan University, Helwan, Egypt, February 2019.
3. Fayek, H.H. Load Frequency Control of a Power System with 100% Renewables. In Proceedings of the 2019 54th International Universities Power Engineering Conference (UPEC), Bucharest, Romania, 3–6 September 2019; pp. 1–6.
4. Rourke, F.; Boyle, F.; Reynolds, A. Tidal energy update 2009. *Appl. Energy* **2010**, *87*, 398–409. [CrossRef]
5. De Almeida, R.G.; Lopes, J.A.P. Participation of Doubly Fed Induction Wind Generators in System Frequency Regulation. *IEEE Trans. Power Syst.* **2007**, *22*, 944–950. [CrossRef]
6. Zaheeruddin; Singh, K. Primary frequency regulation of a microgrid by deloaded tidal turbines. *Soft Comput.* **2020**, *24*, 14667. [CrossRef]
7. Whitby, B.; Ugalde-Loo, C.E. Performance of Pitch and Stall Regulated Tidal Stream Turbines. *IEEE Trans. Sustain. Energy* **2013**, *5*, 64–72. [CrossRef]
8. Bryans, A.; Fox, B.; Crossley, P.; O'Malley, M. Impact of Tidal Generation on Power System Operation in Ireland. *IEEE Trans. Power Syst.* **2005**, *20*, 2034–2040. [CrossRef]

9.   Mauricio, J.M.; Marano, A.; Gomez-Exposito, A.; Ramos, J.L.M. Frequency Regulation Contribution through Variable-Speed Wind Energy Conversion Systems. *IEEE Trans. Power Syst.* **2009**, *24*, 173–180. [CrossRef]

10.   Da, Y.; Khaligh, A. Hybrid offshore wind and tidal turbine energy harvesting system with independently controlled rectifiers. In Proceedings of the 2009 35th Annual Conference of IEEE Industrial Electronics, Porto, Portugal, 3–5 November 2009; pp. 4577–4582.

11.   Rahman, M.L.; Shirai, Y. Hybrid offshore-wind and tidal turbine (HOTT) energy conversion I (6-pulse GTO rectifier and inverter). In Proceedings of the 2008 IEEE International Conference on Sustainable Energy Technologies, Singapore, 24–27 November 2008; pp. 650–655.

12.   Huang, J.; Yang, J.; Xie, D.; Wu, D. Optimal Sliding Mode Chaos Control of Direct-Drive Wave Power Converter. *IEEE Access* **2019**, *7*, 90922–90930. [CrossRef]

13.   Hasanien, H.M. Transient Stability Augmentation of a Wave Energy Conversion System Using a Water Cycle Algorithm-Based Multiobjective Optimal Control Strategy. *IEEE Trans. Ind. Inform.* **2018**, *15*, 3411–3419. [CrossRef]

14.   Ghoshal, S. Optimizations of PID gains by particle swarm optimizations in fuzzy based automatic generation control. *Electr. Power Syst. Res.* **2004**, *72*, 203–212. [CrossRef]

15.   Das, D.C.; Roy, A.; Sinha, N. GA based frequency controller for solar thermal–diesel–wind hybrid energy generation/energy storage system. *Int. J. Electr. Power Energy Syst.* **2012**, *43*, 262–279. [CrossRef]

16.   Rao, R.V.; Savsani, V.J.; Vakharia, D.P. Teaching–learning-based optimization: A novel method for constrained mechanical design optimization problems. *Comput. Des.* **2011**, *43*, 303–315. [CrossRef]

17.   Pan, Q.-K.; Suganthan, P.; Tasgetiren, M.F.; Liang, J. A self-adaptive global best harmony search algorithm for continuous optimization problems. *Appl. Math. Comput.* **2010**, *216*, 830–884. [CrossRef]

18.   Hayyolalam, V.; Kazem, A.A.P. Black Widow Optimization Algorithm: A novel meta-heuristic approach for solving engineering optimization problems. *Eng. Appl. Artif. Intell.* **2020**, *87*, 103249. [CrossRef]

19.   Rawat, S.; Singh, S.; Gaur, K. Load frequency control of a hybrid renewable power system with fuel cell system. In Proceedings of the 2014 6th IEEE Power India International Conference (PIICON), Delhi, India, 5–7 December 2014; pp. 1–6. [CrossRef]

20.   Das, D.C.; Sinha, N.; Roy, A.K. Automatic Generation Control of an Organic Rankine Cycle Solar-Thermal/Wind-Diesel Hybrid Energy System. *Energy Technol.* **2014**, *2*, 721–731. [CrossRef]

21.   Fayek, H.H.; El-Zoghby, H.M.; Ghany, A.A. Design of Robust PID Controllers Using H∞ Technique to Control the Frequency of Wind-Diesel-Hydro Hybrid Power System. In Proceedings of the International Conference on Electrical Engineering, San Francisco, CA, USA, 22–24 October 2014.

22.   Fayek, H.H. Robust Controllers Design of Hybrid System Load Frequency Control. Master's Thesis, Helwan University, Helwan, Egypt, 2014.

23.   Fayek, H.H.; Shenouda, A. Design and Frequency Control of Small Scale Photovoltaic Hydro Pumped Storage System. In Proceedings of the 2019 IEEE 2nd International Conference on Renewable Energy and Power Engineering (REPE), Toronto, ON, Canada, 2–4 November 2019; pp. 32–37.

24.   Kumar, A.; Shankar, G. Quasi-oppositional harmony search algorithm based optimal dynamic load frequency control of a hybrid tidal–diesel power generation system. *IET Gener. Transm. Distrib.* **2018**, *12*, 1099–1108. [CrossRef]

25.   Veronica, A.J.; Kumar, N.S.; Gonzalez-Longatt, F. Design of Load Frequency Control for a Microgrid Using D-partition Method. *Int. J. Emerg. Electr. Power Syst.* **2020**, *21*, 20190175. [CrossRef]

26.   Chen, M.-R.; Zeng, G.-Q.; Dai, Y.-X.; Lu, K.-D.; Bi, D.-Q. Fractional-Order Model Predictive Frequency Control of an Islanded Microgrid. *Energies* **2019**, *12*, 84. [CrossRef]

27.   Hefiri, K.; Garrido, A.J.; Rusu, E.; Bouallègue, S.; Haggège, J.; Garrido, I. Fuzzy Supervision Based-Pitch Angle Control of a Tidal Stream Generator for a Disturbed Tidal Input. *Energies* **2018**, *11*, 2989. [CrossRef]

Publisher's Note: MDPI stays neutral with regard to jurisdictional claims in published maps and institutional affiliations.

 *inventions* 

Article

A New Project for a Much More Diverse Moroccan Strategic Version: The Generalization of Solar Water Heater

Fatima Zohra Gargab [1,2], Amine Allouhi [1], Tarik Kousksou [2], Haytham El-Houari [1,3,*], Abdelmajid Jamil [1] and Ali Benbassou [1]

1 Ecole Supérieure de Technologie de Fes, Université Sidi Mohamed Ben Abdellah USMBA, Route d'Imouzzer, Fès BP 2427, Morocco; fatimazohra.gargab@gmail.com (F.Z.G.); allouhiamine@gmail.com (A.A.); Abdelmajid.jamil@usmba.ac.ma (A.J.); Ali.benbassou@usmba.ac.ma (A.B.)

2 Laboratoire des Sciences de l'Ingénieur Appliquées à la Mécanique et au Génie Electrique (SIAME), Université de Pau et des Pays de l'Adour/E2S UPPA, EA4581, 64000 Pau, France; tkousks@gmail.com

3 Laboratoire de Mathématiques, Modélisation en Physique Appliquée, Ecole Normale Supérieure de Fès, U.S.M.B.A, Route Bensouda, Fez BP 5206, Morocco

* Correspondence: haythamelhouari1@gmail.com

Abstract: This paper presents a strategical project for the new version of the Moroccan energy policy. It highlights the technology of solar water heaters (SWH), studying energy, economic and environmental gains of SWH generalization to satisfy the total resident need proposing a new strategic version diversified in terms of adopted technologies (more than green electricity). A detailed analysis of thermal performances and economic profitability of direct thermosyphon solar water heaters (TSWH) for residential requirements in Morocco. The optimum design parameters were defined and investigated using the dynamic TRNSYS simulation program. The optimum system was simulated under the six climatic conditions of Morocco in order to assess the related performances in terms of the collector efficiency and solar fraction. The major finding of this work is that large-scale integration of TSWH into Moroccan residences could provide up to 70% of thermal energy loads. An economic study was also developed to predict the life-cycle savings generated by the generalization of this technology in Morocco for all residential building's categories. Approximately 1250 million USD as national saving on the total energy bill can be achieved. The environmental effects were also assessed to achieve the aims of this work and to evaluate the CO_2 emissions avoided due to this environmentally friendly solution.

Keywords: solar hot waters; thermosyphon; thermal performance; Morocco; economic outcomes; CO_2 environmental assessment

 check for updates

Citation: Gargab, F.Z.; Allouhi, A.; Kousksou, T.; El-Houari, H.; Jamil, A.; Benbassou, A. A New Project for a Much More Diverse Moroccan Strategic Version: The Generalization of Solar Water Heater. *Inventions* **2021**, *6*, 2. https://dx.doi.org/10.3390/inventions6010002

Received: 16 October 2020
Accepted: 22 December 2020
Published: 24 December 2020

Publisher's Note: MDPI stays neutral with regard to jurisdictional claims in published maps and institutional affiliations.

1. Introduction

Sun is an abundant alternative source of clean, renewable, and sustainable energy [1]. However, it was found that solar energy remains until now not fully exploited because of the high initial cost of solar thermal technologies [2], particularly in Morocco, a member country of the Middle East and North Africa (MENA) and considered as a case study in this work. In fact, one of the most valuable new policies is the integration of solar energy in the building sector to satisfy the needs of domestic hot water in Morocco, principally through solar thermal technologies. Indeed, thermosyphon solar water heaters (TSWH) can reduce intensively the consumption of butane gas, develop the market of solar energy in Morocco, and as a result widen the overall area of installed collectors. The use of such technologies can also participate to reach national targets related to energy efficiency (EE) in the building sector [3]. Finally, solar thermal heaters can enhance the safety aspect during the utilization phase by avoiding the phenomenon of fatal accidents due to the exhaust of butane gas which was estimated at 60 deaths per year according to Gergaud et al. [4].

TSWH has become the most used technology around the world to heat water. It is defined as a passive solar technology based on natural convection, without the necessity of using mechanical or electrical power to pump and circulate the fluid neither its control. Fluid's circulation is the result of the density difference caused by the solar fluid heating considered as a driving force. Thermosyphon is described as the oldest technology which was introduced into the world market. Its performance is equal to the active systems and remains efficient in some cases. Hence, it can be concluded that the thermosyphon technology is the simpler system that could be designed [5].

The performance of solar water heaters has been developed and assessed theoretically and experimentally in several works. For example, Michaelides et al. [6] studied, different configurations of solar water heating systems, especially the under meteorological and socio-economic conditions of Cyprus, using different load profiles where the results showed that the solar fraction balanced between 89 and 63% depending on the consumption pattern studied (low and high). Moreover, Carlsson et al. [7] studied the thermosyphon system by replacing the traditional materials with polymeric ones, in order to compare the thermal, and financial performances. It was found that the total energy costs could be reduced considerably using the polymeric material. Further detailed analyses on Thermosyphon performances were also investigated by Kalogirou et al. [8]. Authors studied the performance of a TSWH considering a series arrangement and formulated a novel heat exchanger model to evaluate the effects of many parameters such as the inclination angle, coupling geometries and aspect ratio on the global performance. The most influential parameter was found to be the aspect ratio which affected to temperature and efficiency according a linear trend [9,10].

Many researchers have focused on innovative techniques to improve the efficiency of TSWH; Vasiliev [11] discussed using of new two-phase termosyphons technologies with advanced performances in order to study identical long construction with different working fluids, to examine the thermal performances especial the resistance which can be lowed using the vapordynamic thermosiphon (VDT) thernosyphons and polymeric loop two-phase under different parameters such us the inclination. The innovation continuous to appear in many other works, such as Piotr Felinski's [12] study which applies the phase change material (PCM) and the paraffin in evacuated tube collectors for domestic hot water application, to evaluate the effect of this materials and thermal performance. The results showed that the use of evacuated tube collector (ETC) with PCM and paraffin allowed height water temperatures and improved solar fractions during the peak loads and the periods of lowest solar radiation intensities under the typical meteorological year conditions, compared to the conventional types. While the conventional ETC showed in several works right performances in terms of water temperatures, yields, energy and exergy efficiencies for many applications; thermosyphonic and forced models [13–15]. Koffi et al. [16] studied theoretically and experimentally a TSWH with an internal exchanger in order to assess the effect of some operating parameters on the outlet temperature and collector thermal efficiency. This study was investigated for Ivory Coast climatic conditions in 2014. Zerrouki et al. [17] studied the outlook for the use of a thermosyphon system manufactured in Algeria to predict its efficiency under the Algerien climatic conditions. Another study, with the same purpose was carried out under clear nights conditions and using the thermosyphon system with at-plat collectors by Runsheng Tang et al. [18]. Many other works were developed recently to study the general performances TSWH. For instance, Zeghib et al. [19] in Algeria presented a new theoretical simulation model, to predict the performance and the system behaviors under a thermosiphonic operation. The model was carried out using a flat plat collector with an area of 2 m^2 for a volume of 200 L. Consequently, the theoretical analysis results height heat energy output, and satisfying efficiencies of collector and auxiliary heater improved by the good stratification of the model designed compared to the fully mixed conventional models. Vieira et al. [20] performed a multiparametric study of SWH system under the climatic conditions of Brisbane in Australia, using EnergyPlus 8.6 software. The model's calculations and analysis showed that the split systems perform better than thermosiphon system in terms of service level and energy efficiency. Another parametric study based on the load profiles influence was carried out by ERICH HAHNE [21]. The analysis was performed using a solar combisystem using a storage tank with an internal thermosiphonally driven discharge unit, operating under realistic profiles of 1-min scales and a constant total yearly demand. The results showed that the load profile have a severe influence on the system performances, this is explained by the significant role of the duration and the flow rates of DHW (Domestic Hot Water) on the stratification and temperatures. S. Fung et al. [22] presented two existing systems in his work, drain water heat recovery (DWHR) system and two solar domestic water heaters (SDWH) intended for two houses, the study aims to evaluate the performances and the recovery potential of systems, using two collector technologies; flat plat collector and ETC as a result the DWHR present an effectiveness of 50%, and the productivity of sensors was evaluated. Several works were investigated to study other types of solar water heaters. The analysis of Hobbi et al. [23] could be cited as example. Authors simulated and optimized a forced circulation using the flat plat collectors for residential SWH application in Montréal Canada, the study aims to show the importance of solar energy which could provide between 83–97% of the hot water demand in summer. The aim is to improve the efficiency of systems integrated in buildings, in order to reach economical and efficient buildings. Something that can be achieved with a multitude of actions [24,25]. Thermosyphon effect has several other technological uses, Chen and Yang [26] solve the efficiency problem for concentrating solar cells due to their height temperature, by the integration of the loop thermosyphon in the heat dissipation system. Different fluid for the operating thermosyphon loop were used. The results of theoretical and experimental analyses showed that the acetone is better than water and ethanol in term of heat transfer. Solomon et al. [27] focused on the closed system presenting the comparison between the two phase closed thermosyphon (TPCT), porous copper coating and the uncoated, the results showed that difference in heat transfer coefficients between the two cases at an angle of $45°$ is 44% at a heat flux of 10 kW.m^{-2}. Furthermore, more the copper coating is thin more the wall temperature of the evaporator is significant and the heat transfer coefficient is increased, which makes the thin coating a suitable for cooling high-density power. Kousksou et al. [28] presented a numerical study which is concerned with the integration of phase change materials (PCMs) in solar-based domestic hot water (DHW) systems, so as to enhance its overall performances.

In the literature, many works join to the technical study an economic analysis to present a significant result, especially for the strategically and policies works [29]. Recent economic studies regarding solar water heaters integration were published. The techno-economic benefits and reliabilities of solar water heaters using the Monte Carlo analysis were estimated by Rezvani et al. [30]. They focused on a product range manufactured by a local company in Australia. Sokka et al. [31] focused on the environmental impacts of solar energy. Greenhouse gas emissions and mitigation of climate change reduction were studied which is the main aim of the global climate policy. Bouhal et al. [32] investigated the impact of collector technology and load profile on the fractional savings of solar domestic water heaters under various climatic conditions in Morocco. While an energy analysis of solar domestic heating water systems was conducted by Allouhi et al. [33].

In the present work, the preliminary choice of the technology to be integrated into the Moroccan context was made based on the results of the different systems studied in the work exhibited in the literature review previously mentioned, the particularity of this study is to integrate the Moroccan climate context, economic and social in the choice and the technical elaboration of the adapted model in order to present a realistic study combining the technology and the current conjuncture of Morocco as part of an African area and fertile ground of the generalization project. For Morocco, solar energy is a crucial economic issue in coherence with the choice of sustainable development and the related energy policy, which aims promoting renewable energies to meet 20% of the country's domestic energy needs, improving energy efficiency to achieve 12% reduction by 2020. Solar energy utilization is expected to play a pivotal role in environment protection and social development [34,35]. This orientation leaded many actions, laws and projects to integrate green solutions precisely the solar within different sectors [36], especially the building sector which presents 20.4% of the national total energy consumption [37], this sector is characterized, firstly by a quick development because of fast population growth and an increasing urbanization rate [38,39].

Through this study, large-scale integration of TSWH into various types of buildings in Morocco is discussed energetically and financially based on climatic data of six different climatic zones of Morocco [35]. The performance indicators describing the system in this work are the solar fraction and collector efficiency. The simulation outputs were then used to develop the economic study, which aims to quantify the possible financial gains of generalizing the utilization of TSWH in the domestic sector of Morocco. This assessment was made by comparing the solar system with conventional gas heaters which is the most used heating option in the country. Two scenarios of subsidies elimination for conventional energy sources are discussed and the national potentiality of CO_2 emissions reduction resulting from generalizing this kind of water heaters is assessed.

2. Moroccan Context Related to the Study Goal

In 2008, Morocco faced many challenges. An energy system marked by an extreme dependence on the outside world, a predominance of petroleum products and hydrocarbons in general, sustained growth in demand, increased rural electrification, and high price and price volatility. Despite the relevance of the 2009 strategy's vision and the break it has initiated, with the desire to increase the share of renewable energy. The strategy made it possible to secure the supply, to initiate the liberalization of the electricity market and position Morocco at the forefront of the climate agenda, however this first version of Moroccan energy strategy have several lacks and brakes, such as the focus on the photovoltaique renewable technology; that requires the adoption of new strategic orientations.

This energy strategy adopted in 2009 has consolidated many achievements and must now be revisited. It is necessary to focus on exploiting this deposit to position the Morocco as the energy transition leader. The exploitation of the Moroccan potential conducted in an integrated and inclusive way should benefit the citizen, the economy, and the State.

Many actions will be taken in the new energy strategy version such as training, research, development and innovation: invest more in human capital to build a pool of business skills and networks of researchers and engineers engaged in a global dynamic, both national and regional. Further, we should promote the initiatives undertaken by the ecosystem approach of "Institut de Recherche en Energie Solaire et Energies Nouvelles" (IRESEN) and its partners (universities). Besides the diversification of the energy technologies adopted (more than photovoltaiques and electricity production) is one of the necessaries actions. In this sense, Morocco lunched several projects to support the solar thermal technologies and specially for residential uses. The PROMASOL (programme de développement du marché marocain des chauffe-eau solaires) project was one of the first lunched since the first version which aims to develop the Moroccan solar thermal market, but it didn't make significant results. For this reason, Morocco lunched a new project about solar water heaters which aims to present a new Moroccan SWH with an accessible price and adapted technology to the economic and the climatic context.

For that, an overview of the solar thermal market in morocco is necessary to introduce the present study, and situate this project proposition of (generalization of SWH).

Morocco's equipment level has always remained around four times lower than the global average; this means, notably, its equipment level remained parallel to the world's equipment level. Despite its good sunshine, Morocco is still not among the countries installed capacity in 2018 but, Figure 1 shows that it is listed in the third group of countries (total of five), having between 20 and 70 m^2 for 1000 habitants [40].

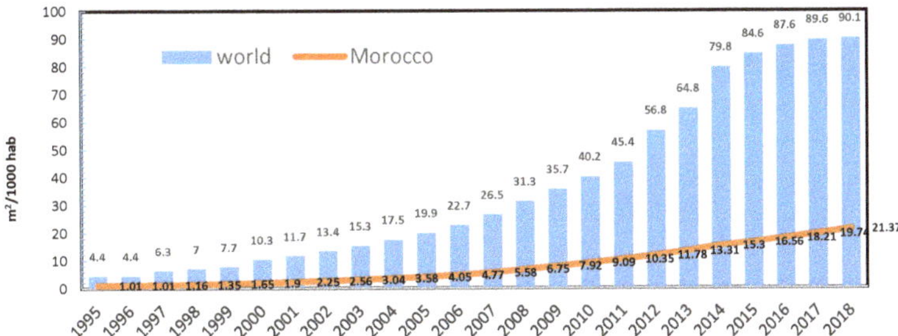

Figure 1. Evolution of the total installed area per inhabitant in Morocco compared to the world.

Morocco's share of installed solar thermal capacity is low but remained stable for three decades as showed in Figure 2, around 0.10%. This means that the growth in m^2 installed in Morocco has been parallel to the world trend [41].

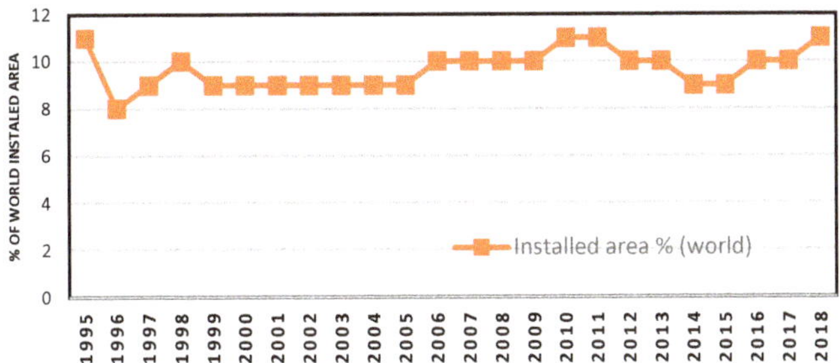

Figure 2. Evolution of the accumulation of the installed area in Morocco (% from the total in world).

On a 30-year scale, Figure 3 shows that: prices in Dh/m^2, cost insurance and freight (CIF, blue diamonds in Figure 3), fluctuated by 15% (vertical blue bars in Figure 3) around a prolonged decrease of nearly 0.34% per year (decreasing blue curve in Figure 3). Solar thermal is no longer in the maturation phase and no longer undergoes a "learning curve." Thus, prices would have lost only 5 to 6% in 30 years, which is low but appreciable considering inflation and increased purchasing power. The thermal energy produced by solar collectors has always been more profitable than that produced by grid electricity. Even very modest, this drop in prices combined with the rise in electricity prices has made this profitability even more attractive. Although "fair play," this competitiveness goes unnoticed because of the formidable competitor of thermal energy produced by butane gas, which is heavily subsidized. Indeed, the withdrawal of the compensation 12 kg bottle would place its price around 120–140 MAD (Moroccan dirham) and not 40 MAD. Because of this, in Morocco, a solar water heater depreciates in 10 years against its butane gas competitor, regardless of the advantages related to comfort and safety of use.

The installation of solar thermal systems would have contributed to creating a few dozen jobs until around 2007. The number of jobs created amounted to a few hundred until around 2018 and seemed to stagnate around a little over 200 [42].

The annual demand has almost multiplied by 15 in 25 years or average growth of nearly 11% per year, which is not very fast compared to the growth of other renewable energies such as photovoltaics [43]. It is even surprising that the demand is maintained over such a long period despite the harsh competition of butane gas.

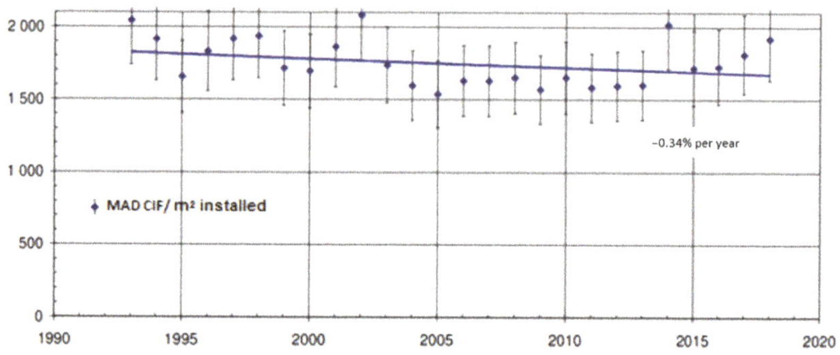

Figure 3. Evolution of the unit wholesale price of thermal collectors (MAD per m^2) [42].

Domestic Solar water heaters (DSWH) have always dominated the sales of the market as described in Figure 4, making a significant turnover presented in Figure 5. As showed in Figure 6 they tend to dominate it more and more to exceed 80% after having fallen around 75% in the middle of the decade 2000–2010, probably under the PROMASOL Program's impetus, which was led by AMEE (Agency for Control and Energy Efficiency).

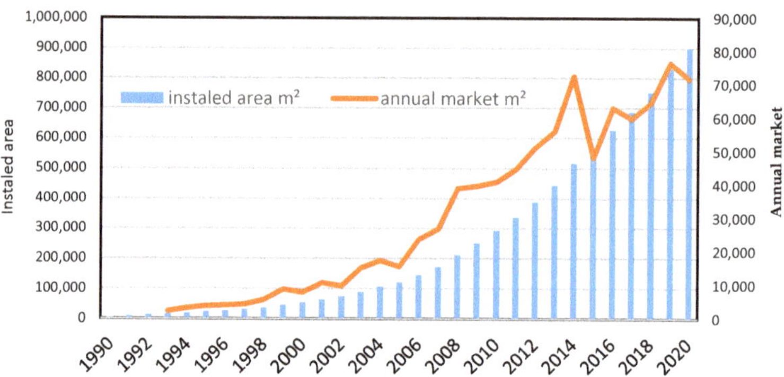

Figure 4. Evolution of the solar thermal collector's cumulative area installed annually in Morocco vs. the national market (imported: installed and not installed).

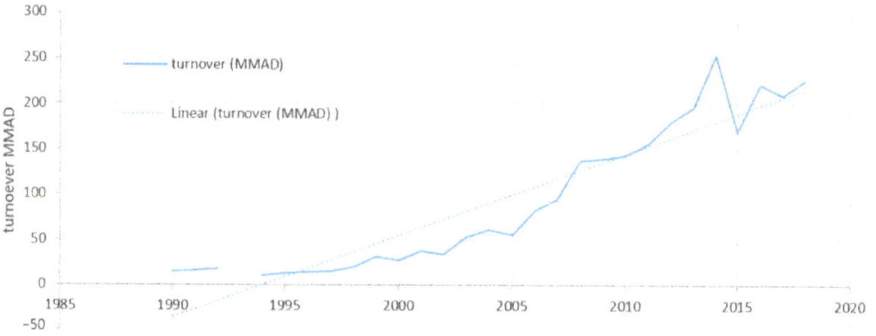

Figure 5. Turnover evolution (solar thermal Moroccan market) for 3500 MAD/m^2. [42].

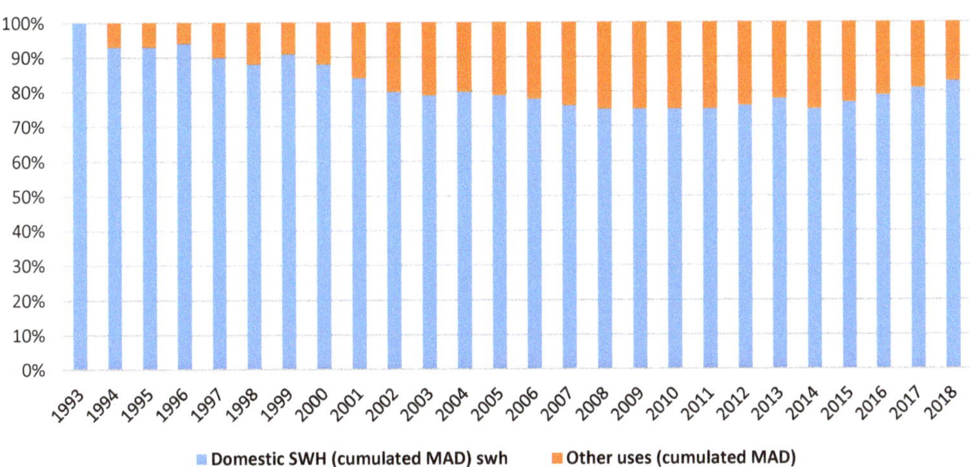

Figure 6. Domestic solar water heaters sales in Morocco.

All the elements presented describe the Moroccan market as a fertile ground for the insertion of this technology and popularizing it. This study assesses the positive impacts of the integration and generalization of this technology at the national level in order to improve the previously presented results further and generate a significant share of the economic and environmental gains not only by the production of green electricity but also by the production of domestic hot water and the coverage of all residential needs by solar.

3. Methodology

The TSWH is simulated using the TRNSYS (transient system simulation tool) program. The main operating and design parameters were introduced; the design parameters include general specifications of the used thermal collector, storage tank and auxiliary system. Weather data files for the various Moroccan climatic zones are then integrated into the simulation process. A realistic load profile is as well considered for accurate prediction of thermal loads and system performances. Based on the hourly output data, namely water outlet in the solar collector and temperature inside the tank, energetic performance indexes were built. These indices are the monthly solar fractions and collector thermal efficiencies.

Based on the evaluated useful energy produced by the TSWH and the total thermal loads required, a large-scale economic study is developed to examine the potential of generalizing these systems into the Moroccan residential sector. This evaluation is based on a comparison of cumulative energy savings generated by switching from the conventional gas heaters to TSWH. Finally, it was possible to estimate the overall reductions in CO_2 emissions.

The assessment is believed to be sufficiently accurate as it follows a rational methodology for quantifying the full-scale economic impact of generalizing these solar systems. First of all, it was necessary to evaluate the number of capita per climatic zone, this parameter is not yet available, it is available only in a regional database according to the High Commission of Plans. Therefore, it was essential to convert the available regional data to comply with the climatic zoning. A schematic diagram of the followed methodology is shown in Figure 7.

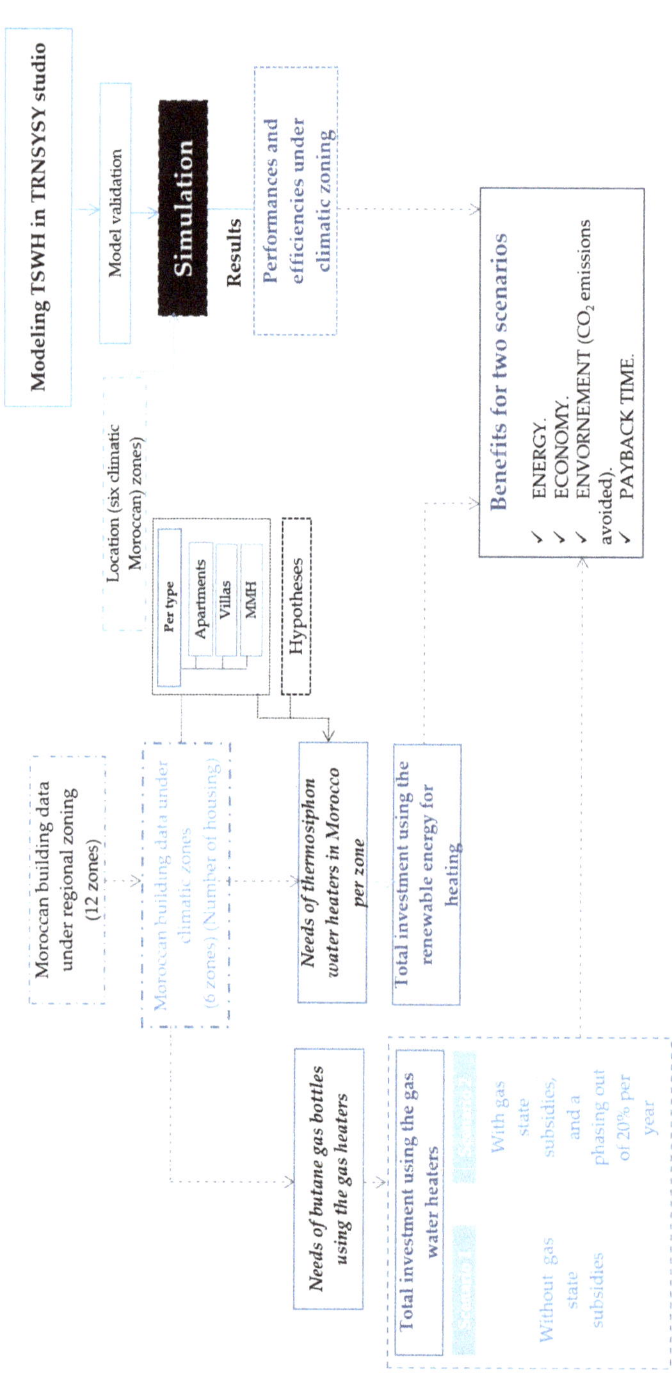

Figure 7. Methodology.

4. Simulation Process

4.1. Examined Configuration

The TSWH consists of a flat plate collector (FPC) connected to an auxiliary heater integrated inside a horizontal tank as presented in Figure 8. For this system, water flows through the system when warm water rises as cooler water sinks without the need for a circulation pump. The FPC collector is installed below the storage tank so that warm water will rise into the tank by the thermosiphon effect, its principle operates on the heated water's characteristic, which becomes lighter than the cold water, ascends to the horizontal tank, replaces the cold water, which, heavier, descends downwards and passes through the solar panel.

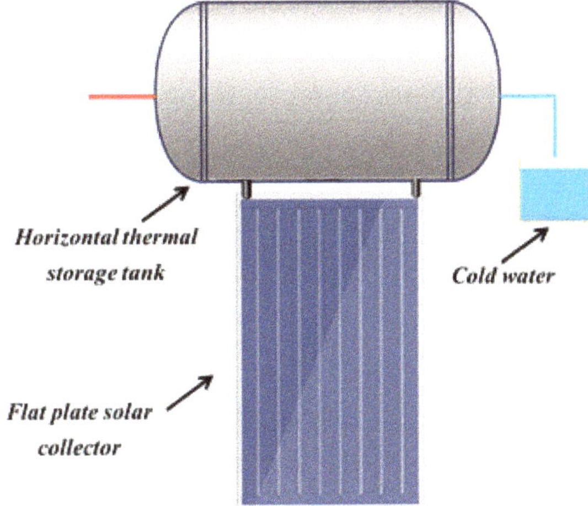

Figure 8. Thermosyphon system schematic.

First of all, optimum tilt angles per each zone were determined. These angles are summarized in Table 1.

Table 1. Optimal tilt angle.

Zone	Agadir	Marrakech	Errachidia	FES	Ifrane	Tangier
Optimal angle	30°	31°	31°	32°	30°	32°

Moreover, the technical characteristics of the solar collector used during the simulations under TRNSYS are presented in the Table 2.

Table 2. Thermosyphon solar water heater characteristics.

Parameter	Value	Unit
Collector area	2.2	m^2
Intercept efficiency	0.74	-
Efficiency slope	12.33	$kJ \cdot h^{-1} \cdot m^{-2} \cdot K^{-1}$
Tested flow rate	30	$kg \cdot h^{-1} \cdot m^{-2}$
Collector slope	30–32	°
Number of parallel collector risers	10	-
Riser diameter	0.2	mm
Header diameter	0.2	mm
Header length	1.0	mm

Table 2. *Cont.*

Parameter	Value	Unit
Collector inlet to outlet distance	1.273	mm
Collector inlet to tank outlet distance	0.25	mm
Collector inlet diameter	0.2	mm
Length of collector inlet	1.0	mm
Collector outlet diameter	0.2	mm
Length of collector outlet	1.0	mm
Outlet pipe losses coefficient	15	$kJ \cdot h^{-1} \cdot m^{-2} \cdot K^{-1}$
Tank volume	202	L
Fluid specific heat	4.190	$kJ \cdot kg^{-1} \cdot K^{-1}$
Fluid density	1000.0	$kg \cdot m^{-3}$
Tank configuration	HORIZENTAL	-
Maximum heating rate	2	kW
Maximum pressure	8	Bars
Thermal losses of the tank	1.8	$kW \cdot h^{-1} \cdot m^{-2}$
Capacity of solar circuit	5	L

4.2. Climatic Data and Geographical Specification

The Moroccan climatic conditions used to assess the thermal performance of the thermosyphon are presented in Table 3 and Figure 9. This new climatic zoning is established by the Moroccan Agency of Energy Efficiency (AMEE).

Because the Moroccan climate exhibits a great variability according to the geographical zone, it was necessary to perform a segmentation of the whole territory into areas with similar overall meteorological specifications. Accordingly, the Moroccan Agency of Energy Efficiency (AMEE) [44], proposed a climatic fragmentation of Morocco according to six climates, each climate is evidenced by a representative city. Table 2 shows general information about these zones. In addition, the main meteorological data (i.e., monthly average ambient temperature and solar irradiations) affecting the energetic performance of TSWHs is displayed in Figure 9. One can observe that the climate presents a remarkable variety which is expected to influence greatly the overall thermal behavior and performance of TSWHs.

4.3. Load Profile

The distribution of the hourly hot water consumption is affected by different factors. It varies according to the seasons, days, or the habits of each considered Moroccan family. Various hot water load profile has been studied on the literature review such as the constant, the early morning, the late morning, rand, or the late afternoon profiles. For the present study, the consumption of 202 L per day is described according to the realistic load profile presented in the Figure 10.

Table 3. General Information of the climatic zones under investigation.

Climatic Zone	Representative City	Altitude (m)	Coordinates	Climate	Temperature °C
Z1	AGADIR	31	30°25′ N 9°36′ W	Subtropical-semiarid	14.3–23.2
Z2	TANGER	20	35°46′ N 5°48′ W	Mediterranean hot	12.6–24.2
Z3	FES	403	34°03′ N 4°58′ W	Mediterranean/continental	9.6–26
Z4	IFRANE	2019	33°32′ N 4°58′ W	Humid temperate climate	6.2–25.4
Z5	MARRAKECH	457	31°37′ N 8°00′ W	Semiarid	12.4–28
Z6	ER-RACHIDIA	141	31°55′ N 4°25′ W	Hot desert	9.1–33.2

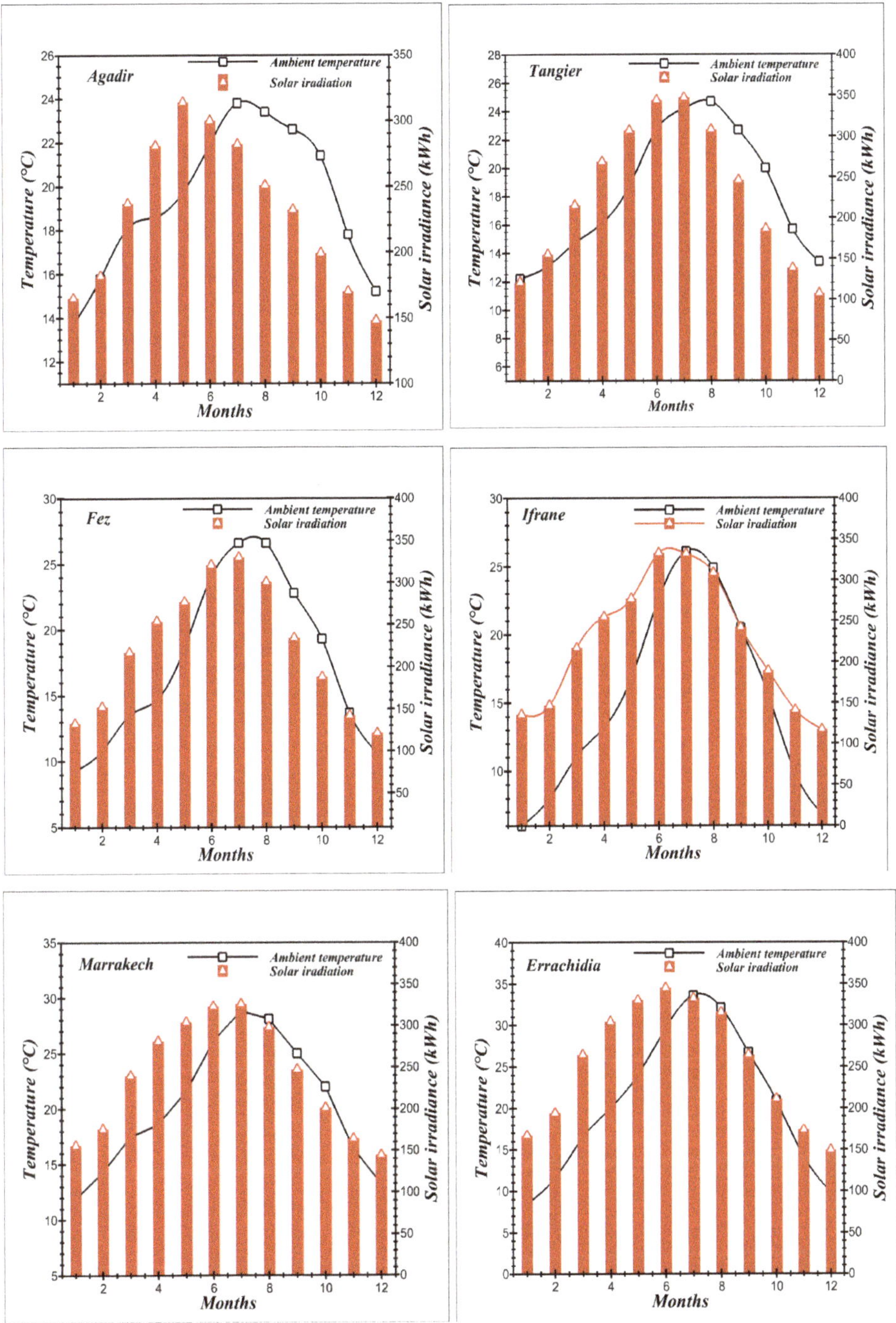

Figure 9. Weather data.

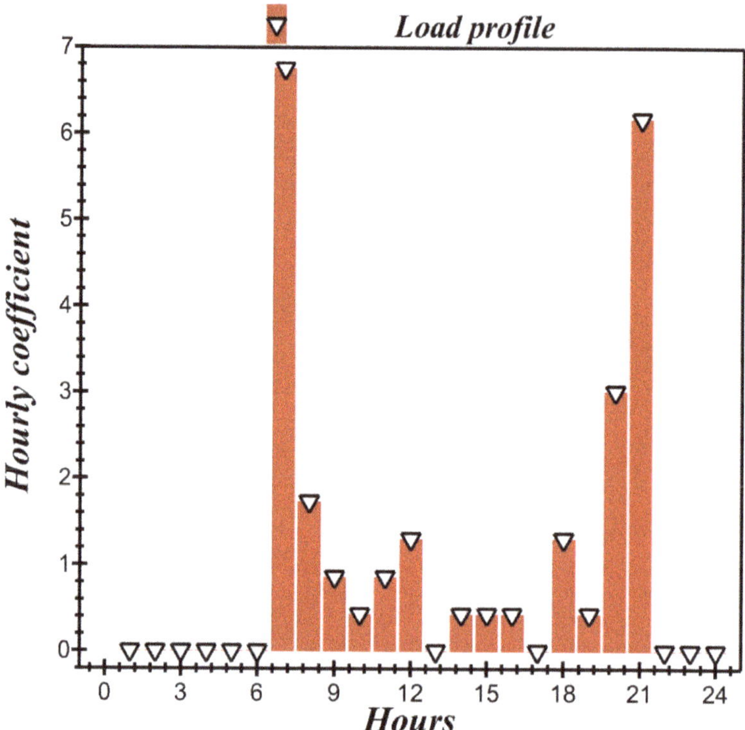

Figure 10. Realistic Moroccan load profile for hot water consumption.

4.4. Physical Modeling

The type 45 operates according to a mathematical model [45] which presents the various phenomena describing its operation. The Bernoulli formula can be applied to define the pressure drop in the thermosyphon system's nodes (Equation (1)):

$$\Delta P_i = \rho_i g \Delta h_i + \rho_i g h_{Li} \tag{1}$$

where (i) is the index of the node, Δh_i the height of the ith node, ρ is the density, h_{Li} is the frictional head loss in the piping, and g is the gravitational constant.

The flow rate on the thermosyphon system must be involved in order to satisfy the conditions describing the total pressure differences, which are determined using the equation above at any time by the next expression (2):

$$\sum_{i=1}^{i=N} \rho_i \Delta h_i = \sum_{i=1}^{i=N} \rho_i h_{Li} \tag{2}$$

Many types of losses are present in the system; the most important one is the frictional head loses especially on the pipes, and the equation below describes this type of losses (3):

$$H_p = \frac{fLv^2}{2d} + \frac{Kv^2}{2} \tag{3}$$

The parameter F is the friction factor, determined according to the value of the Reynolds coefficient R_e which describe the nature of the flow and corrected to take in account the frictions on the connecting parts of pipes. The head loses pipe varies according to the position and the conditions in the system, these specifications are taken into account by the coefficient K, e.g., loses due to different types of bends, the parts of the tank which are connected to the collector.

The thermal performance of collector is modeled according to the Hottel–Whillier equation. Many parameters describe the system: the equation bellow (4) defines the $F'U_L$ parameter which is calculated by the $F_R U_L$ and G parameters determined on the test conditions.

$$F'U_L = -G_{test} \, C_P \ln\left(1 - \frac{F_R U_L}{G_{test} C_p}\right) \tag{4}$$

For the collector the temperature in the K_{th} node is calculated by the next expression (5):

$$T_{ck} = T_a + \frac{I_T F_{R(\tau\alpha)}}{F_R U_L} + \left(T_{ci} - T_a - \frac{I_T F_{R(\tau\alpha)}}{F_R U_L}\right) \exp\left[\frac{F'U_L}{GC_p} \frac{(k - 1/2)}{N_x}\right] \tag{5}$$

where N_x is the number of nodes over which the size of the collector deviates to obtain the weight of the fluid in the collector.

The changes in the fluid heat transfer coefficient results in F' and U_L which are considered negligible. The combination of the intercept efficiency at normal incidence F_R $(\tau\alpha)_n$, and the incidence angle modifier (IAM).

$((\tau\alpha))_n$ allows us to calculate the parameter $F_{R(\tau\alpha)}$ (6):

$$\frac{(\tau\alpha)}{(\tau\alpha)_n} = \frac{I_{bT}\frac{(\tau\alpha)_b}{(\tau\alpha)_n} + I_d \frac{1+\cos\beta}{2}\frac{(\tau\alpha)_s}{(\tau\alpha)_n} + I_g \frac{1+\cos\beta}{2}\frac{(\tau\alpha)_g}{(\tau\alpha)_n}}{I_T} \tag{6}$$

where the incidence angle modifiers for ground and sky diffuse radiation is calculated using the next Equation (7):

$$\frac{(\tau\alpha)_b}{(\tau\alpha)_n} = 1 - b_0\left(\frac{1}{\cos\theta} - 1\right) \tag{7}$$

The overall useful energy collected is determined using the Equations (8) and (9):

$$Q_u = r \, A_c\left(F_{R(\tau\alpha)} \, I_T - F_R U_L(T_{ci} - T_a)\right) \tag{8}$$

$$r = \frac{F_{R(use)}}{F_{R(test)}} = \frac{G\left(1 - \exp\left(-\frac{F'U_L}{GC_p}\right)\right)}{G_{test}\left(1 - \exp\left(-\frac{F'U_L}{G_{test}C_p}\right)\right)} \tag{9}$$

The outlet temperature from the flat plat collector is finally deduced as (10):

$$T_{CO} = \frac{Q_u}{\dot{m}C_p} + T_{Ci} \tag{10}$$

The thermosyphon diagram executed under TRNSYS program is shown in Figure 11. The different module's types used in this study are presented as follows:

- Weather data (TYPE109-TMY2) this component reads meteorological data generated by meteonorm software and supply them in the TMY2 format. It calculates the necessary solar radiation for the calculation of TRNSYS software at any surface tilt.
- Differential controller for temperature (TYPE 2). This component monitors and controls the average temperature of the tank, to control the operation of the auxiliary electric heater.
- The general forcing function (TYPE 14) from TRNSYS library, this component describes and characterizes the hourly load profile corresponding to 202 L/day of hot water demand.

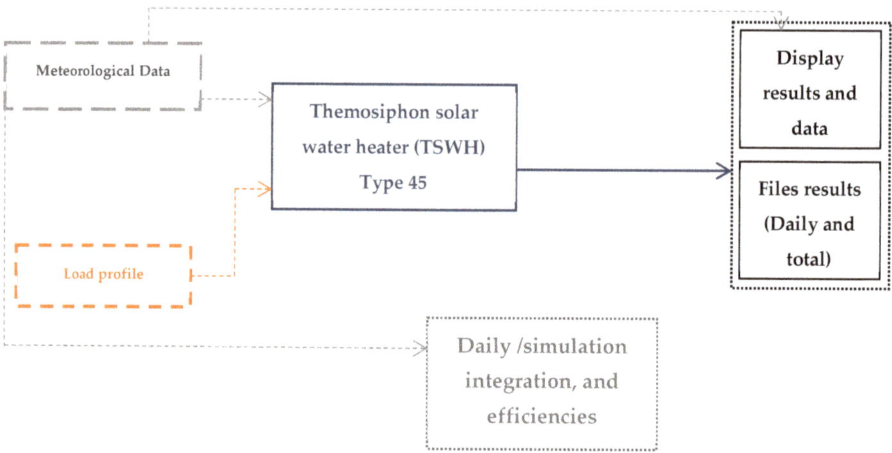

Figure 11. Thermosyphon diagram executed under the TRNSYS program.

4.5. Performance Indices

4.5.1. Collector Efficiency

The collector efficiency describes the ratio of the total useful energy gain to the solar energy absorbed by the collector. It is calculated using the equation below (11):

$$E_{coll} = Q_{ucoll} / (A_c\, I_{coll}) \tag{11}$$

Q_{ucoll} is the energy rate from heat source, and I_{coll} is the total radiation on the tilted surface.

4.5.2. Solar Fraction

The solar fraction F_{sol} represents the solar energy available to meet the needs. It presents a clear contribution indication of the solar system in meeting the thermal load. The solar fraction is calculated using the equation below (12):

$$F_{sol} = 1 - (Q_{aux} / Q_{DHW}) \tag{12}$$

where Q_{DHW} is the energy rate delivered to the load describing the necessary energy to meet all domestic hot water needs, and Q_{aux} is the auxiliary energy. It is important to state that the solar fraction is evaluated on a monthly basis.

5. Model Validation

The developed model was simulated with firstly the meteorological data of Cyprus in order to carry out a comparison with the study performed by Michaelides et al. study [6]. A typical meteorological year (TMY) for Nosica was built for this end.

The system simulation was lunched for similar conditions as indicated in [6]. Table 4 shows the technical specifications of the studied TSWH while Tables 4 and 5 indicates the general assumptions concerning technical parameters of validation case, the load profile and water consumption.

Table 4. Consumption profile of the validation study.

Profile Code	Daily HW Consumption L/Person	Temperature (°C)	Pattern
DOM	30	50	Michaelides

Table 5. Technical parameters of the validation study.

Parameter	Value
A_c	2.72 m^2
F_R	0.791
$F_R U_L$	24 kJ·h^{-1}·m^{-2}·k^{-1}
G_{test}	96 kg·h^{-1}·m^{-2}
β	40°
N_R	20
D_r	15 mm
d_i, d_o	22 mm, 22 mm
H_c	1000 mm
H_O	1150 mm
H_r	250 mm
H_a	300 mm
H_{th}	370 mm
L_i, L_o	2000 mm, 520 mm
U_{api}	5.1 kJ·h^{-1}·k^{-1}
U_{APo}	5.1 kJ·h^{-1}·k^{-1}
V	162 L
$(UA)_s$	6.45 kJ·h^{-1}·k^{-1}
Q_{aux}	3 kW

Figure 12 presents a comparison between monthly solar fractions of the TSWH investigated by Michaelides et al. study [6] and those predicted by the introduced TRNSYS model.

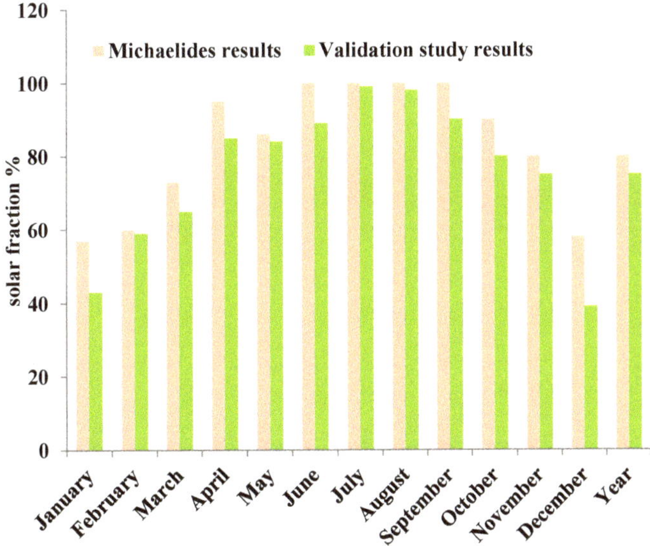

Figure 12. Solar fraction of Michaelides study versus validation study.

The average relative error is about 5%, which is an acceptable value. The deviation between the results can be attributed to the update in climatic conditions caused when using the Meteonorm database that corresponded to the period 1986–1992 for the study of Michaelides et al. [6] while the prepared metorological data file used in our calculations corresponds to the period 1991–2010.

6. Results and Discussion

6.1. Energetic Evaluation

As presented in Figure 13a, the thermosyphon system performs differently in the six cases. The average solar fraction ranges between 46% to 71%, and the maximum solar fraction value was noticed in Errachidia 71%. This result is explained by the importance of solar irradiations in this zone. The annual average solar fraction is about 69% in the first zone represented by Agadir which is a sunny coastal zone. It is in general the most favorable zone for this system, especially in winter.

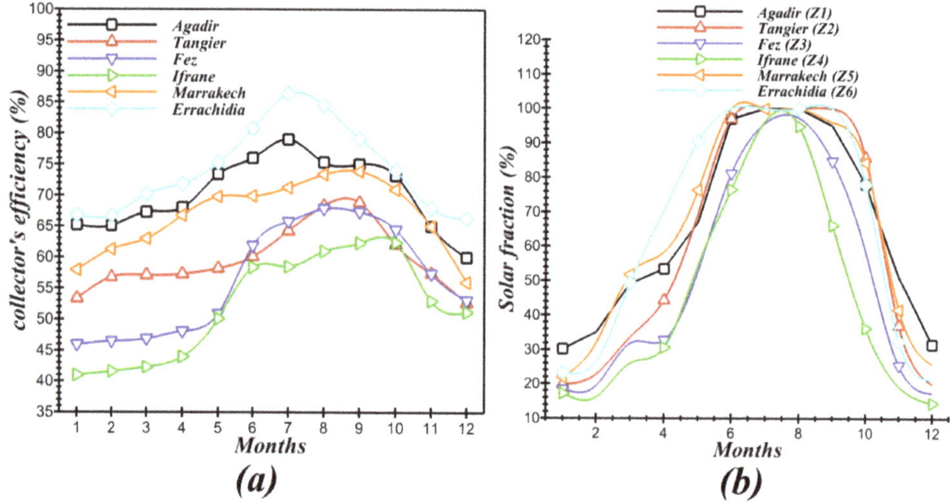

Figure 13. (**a**) Solar fraction. (**b**) Collector's efficiency.

Tangier (Zone 2) and Marrakech (Zone 5) exhibit approximately the same solar fraction variation, but (Zone 5) has a higher annual average value of 66% versus 61% in (Zone 2). This difference is due to the favorable conditions of Marrakech which is considered as one of the sunniest areas in Morocco. Zone 3 presents a total annual average solar fraction of 52%, which is lower compared to the previously discussed sites. This result can be explained by the fact that Fez is characterized by a Mediterranean climate, dominated by continental and Atlantic effects, which accounts for the remarkable fall of the solar fraction to 17.4% during the coldest month of winter. Finally, the coldest zone, Ifrane has an averaged solar fraction value of only 46%. It is concluded that Ifrane's weather is the most unfavorable climate for installing solar TSWHs because Ifrane remains the coldest zone in Morocco with an average annual temperature of 10.8 °C. In addition, one can observe that most of the hot water requirements in the Moroccan zones, previously cited, are totally satisfied during the sunniest three months of the year (July, August, and September) with a solar fraction achieving 100%.

Figure 13b presents the dynamic behavior of the collector efficiency for the six zones. The most favorable zone, where the collector performs with high efficiency is Zone 6 represented by Errachidia, with a maximum value of 86% in July, and an average annual value of 74%. The collector operates in Agadir with 70% of efficiency, in Marrakech with 66%, Fez and Tangier with 54% and 59% respectively and finally in Ifrane with 51%. These results confirm the explanations previously presented for solar fraction. In fact, when the thermal efficiency of the collector is maximized, the useful thermal energy transmitted to the working fluid is as well maximized and thus less auxiliary energy will be required.

6.2. Economic Assessment

This section aims to assess the total financial gains generated by the integration and the generalization of solar water heaters in Morocco into the residential sector, according to the six climatic zones and for different house's types. The solar system will be compared to a reference system which is the conventional gas boiler according to two scenarios:

1. The first scenario takes into account the state subsidies on gas, which presents a purchasing power support of households. In this case, the consumer benefits on each bottle of gas that purchased a reduction of 67% on its price.
2. The second scenario does not consider the subsidies.

The number of thermosyphon systems required to be installed was determined according to the situation of the building in Morocco. The number of houses counted on 2014 is 5.8 million, distributed as follow: occupied houses, secondary, vacant and professional use houses as presented in Table 6.

Table 6. Distribution of houses in Morocco according to the 2014 census.

Type	Percentage
Housing occupied (habitat)	79%
Secondary housing	6%
Vacant	9%
Professional use	3%

The number of occupied houses in Morocco is 4.582 million houses: this value is officially distributed over 12 regional zones according to (the High Commissariat of the Moroccan Plan) [34].

A conversion of this distribution according to the climatic zones was investigated as presented in Table 7. It is found that this number of houses per climatic zones can be also fragmented into many types such as: modern Moroccan houses which represent the main construction type in Morocco with a percentage of 63%, followed by apartments (25%), villas 4%, traditional Moroccan houses 4%, and finally anarchic construction 4%.

Table 7. Distribution of occupied houses per climatic zone in Morocco.

Zone	Percentage of Houses %	Number of Houses
1	46	2,109,220
2	12.5	572,790
3	29.3	1,340,970
4	3.1	144,960
5	4.1	189,740
6	5	224,320

These statistics will enable the determination of the thermosyphon's number to be integrated at a national scale.

The study will focus on apartments, modern Moroccan houses, and villas. It should be highlighted that, because of the presence of these variants in the construction sector, it is recommended to consider different capacities related to the type of house for an appropriate analysis. Table 8 presents the system capacity concerning the construction type. The generalization of this system into the building sector will generate an important total investment of 5876.42 million USD. This total investment is found to be 1145.5 million USD, 4144.42 million USD, and 586.50 million USD for modern Moroccan houses (150 L), apartments (200 L), and villas (300 L) respectively.

Table 8. The number of thermosyphon solar water heaters (TSWH) to be installed.

Size	150 L		200 L		300 L	
Zone	Number	Area m^2	Number	Area m^2	Number	Area m^2
Zone 1	527,400	1,160,100	1,413,200	3,109,000	84,400	371,300
Zone 2	143,200	315,100	3,837,800	844,300	23,000	100,800
Zone 3	335,300	737,600	898,500	1,976,600	53,700	236,000
Zone 4	36,300	79,800	97,200	213,700	5800	25,600
Zone 5	47,500	104,400	127,200	279,700	7600	33,400
Zone 6	56,000	123,400	150,300	330,700	8900	39,500

6.3. Total Annual Gains

The total gains or the life cycle savings of the thermosyphon generalization in Morocco by using the subsidies and non-subsidies gas butane prices is presented in Figure 7. This indicator represents the global cost savings resulting from the integration of the thermosyphon system instead of buying gas butane cylinders.

It is concluded, from the comparison of these two scenarios, that, obviously, the gains generated considering subsidies are less than the second case for the six zones. In fact, the government subsidies (on fuels) policy is representing a barrier towards development solar energy projects in Morocco. Moreover, the first zone represents the most important gains by 197.11 million USD in the first scenario against only 44.32 million USD in the second zone. This difference can be explained by the highest solar fraction (68.8%), together with the high intensity of the population concentration with a percentage of 46%. In the zone represented by Agadir, this optimal combination of solar fraction and the number of homes for the first climatic zone gives optimized results and very encouraging gains as it is previously represented. This value becomes significant and would reach 448 million USD per year by canceling government grants.

The third zone is holding the second place with total gains of 69.34 million USD per year. It can be improved by removing the state subsidies to reach 214.63 million USD per year. The other climatic zones achieve lower gains than the previous ones, despite their significant solar fractions which reach 66% (Errachidia) representing the most favorable climate conditions for the system. This result is explained by the minimal percentage of houses.

The energy gains generated by the generalization of thermosyphon systems are presented in Figure 14b. The first zone achieve an energy annual gain of 5943.46 GWh per year, which is distributed according to the different studied configurations: apartment with 1214.60 GWh per year, modern Moroccan houses by 4340.19 GWh annually, and 388.67 GWh generated for the villas type. The integration of solar water heaters in the modern Moroccan houses generates the maximum gains in the six climate zones compared to other types due to the remarkable presence of this type in the typological distribution of housing in the Moroccan territory. In fact, it represents 63% of the total occupied houses, followed by the buildings and finally villas.

Figure 15a–c present the evolution of the annual gains in an interval of 12 years, and under the condition of (non-subsidies) for several type of houses treated in this paper.

The payback period for the proposed sizes and under the six climatic data condition can be determined based on Figure 14. The return on investment (ROI) varies between 3 and 6 years according to the performances of the system related to the meteorological context, the investment and the gains generated which are related also to the performance and the total expenditures produced by the traditional gas butane system. In general, after a period of 6 years the thermosyphons installed in the climatic zones will cover the initial investment and in general they will enable 100% of the life cycle savings.

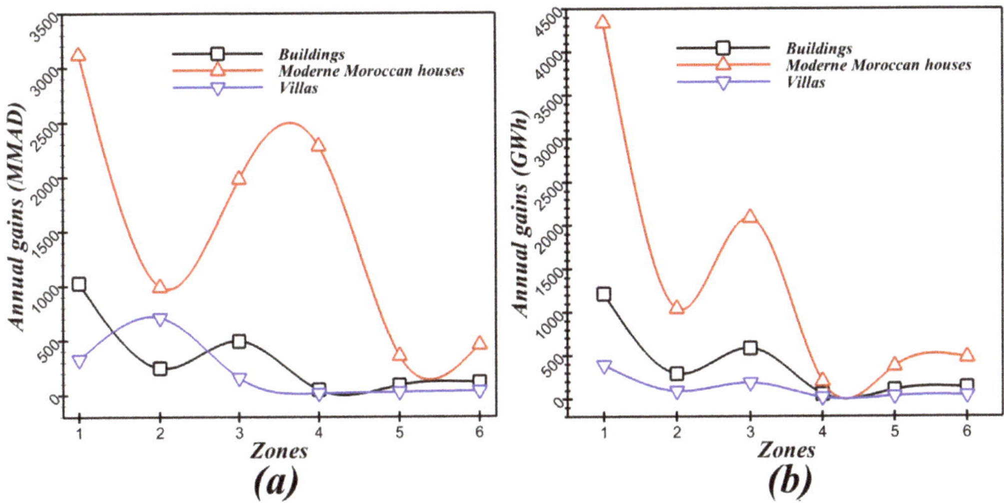

Figure 14. (**a**) Total annual gains (with state subsidies). (**b**) Total annual gains (without state subsidies).

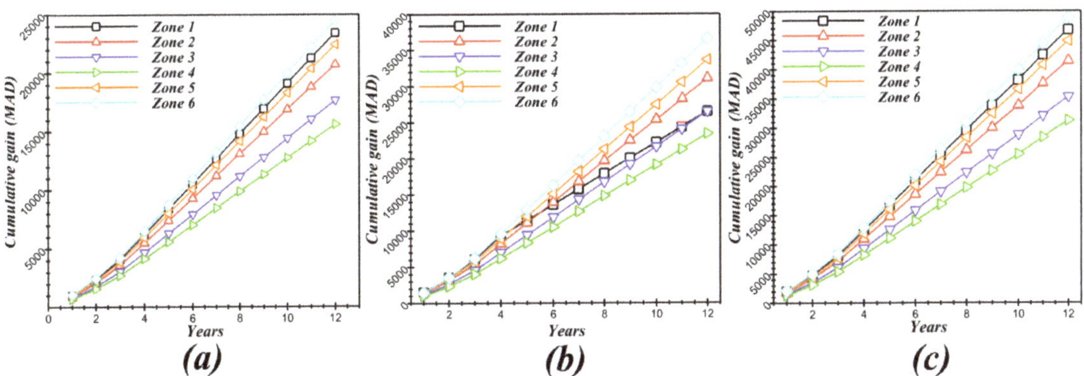

Figure 15. (**a**) Cumulative gains for buildings (non-subsidies scenario). (**b**) Cumulative gains for modern Moroccan houses (non-subsidies scenario). (**c**) Cumulative gains for villas (non-subsidies scenario).

Global Results: Cumulative Gains for Different House's Types

The overall average solar fraction presents a national value which is estimated at 62%. Table 9 presents the total gains for different type of house according to the two scenarios (with and without subsidies): the generalization of solar water heaters have an important role to play in the achievement of the development program's goals of the Moroccan market for solar water heaters by (PROMASOL).

This program aims to provide the evolution continuity of thermosyphon installed capacity since 2000 obtained by the analysis of imports conducted by the (AMEE) which is a result presented in Figure 16. This figure is presenting the cumulative area of the SWH installed in Morocco since 2000 to the expectations desired to achieve.

Table 9. The total gains for different type of houses according to the two scenarios (with and without subsidies).

Type of House	Apartments	Modern and Traditional Moroccan Houses	Villas
Total Investment	1145.5	4144.42	586.4
Subsidies			
Annual expenditure of gas Water heater (USD)	115.47	3009.45	36.95
Gains (USD)	71.59	1865.86	22.91
No subsidies			
Annual expenditure of gas Water heater (USD)	357.40	957.83	114.36
Gains (USD)	221.59	593.86	70.90
The annual gains (the number of gas bottles)	17.05 million	45.68 million	5.46 million

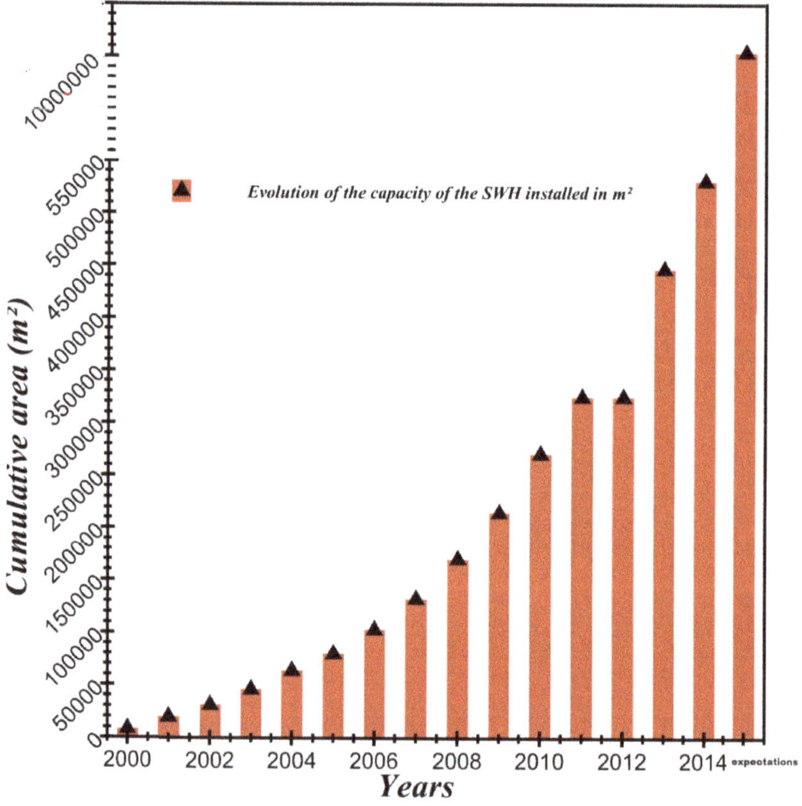

Figure 16. Cumulative area of the SWH installed in Morocco since 2000 and expectations.

6.4. Environmental Assessment

Using solar water heaters will also make it possible, in contrast to the conventional water heater, to reduce emissions of greenhouse gases. It should be recalled that recently, Morocco announced its engagement to reduce 13% of these gases by 2030, and ratified the Kyoto Protocol, which entered into force in 2005 Table 10 presents the annual CO_2 emissions avoided by the integration and the generalization of the TSWH.

Table 10. The annual CO_2 emissions avoided by the integration and the generalization of the TSWH.

Zone	Gains (Number of Gas Cylinder) (million)	Gains (million m^3 of Butane Gas)	CO_2 Emissions Avoided (tones/year)
Zone1	44,610.75	220,377	1,856,000
Zone 2	8.36	41.29	446,900
Zone 3	16.51	81.56	891,800
Zone 4	1.59	7.86	85,300
Zone 5	3.1	14.83	160,200
Zone 6	3.56	1.756	203,700

The maximum CO_2 emissions avoided are remarkably visualized on the first climatic zone which presents the highest number of houses, and as a result the highest number of SWH installed, and energy savings.

The calculation of the CO_2 emissions avoided is made based on the quantity of the butane gas consumed in the conventional system; this consumption of the butane gas will be eliminated by implementing the generalization project. This step required adopting the butane gas characteristics where it generates 230 g of CO_2 emissions for each kWh produced. Furthermore, for bottles of 6 kg, the discharges correspond to 157 tons of CO_2 equivalent. The bottles of 10 kg, under the same conditions, give 160 kg of CO_2 equivalent. They finally amount to a bottle of 13 kg to 150 kg of CO_2 equivalent.

7. SWH Generalization a Bridge Idea to a Promoter Economic Project (Manufacturing Moroccan SWH)

Nowadays, the global perspective focuses on the economy, the human, energy, and the environment, achieving optimum energy consumption and reducing greenhouse gas emissions. Besides, with this 2020 pandemic, the economic sector's growth has become more urgent than before.

In Morocco, the project of this paper can be realized currently only by the import of this technology; this requires reflection to a local production unit of solar water heaters. Here, the idea of an innovative project was born (SOL'R SHEMSY project). We are working through the development of a Moroccan solar water heater ready to be industrialized locally. Hence, the project integrates all these goals and offers a balanced solution; technology, economy, and social and environment.

The local production will present solar water heaters with affordable prices adapted to the social context, breaking the most significant barrier of this technology's extensive integration, which is the high prices resulting from the importation effect.

The manufacturing project will bring various high-quality products intended for multiple customer categories to meet all social and industrial clients' needs, presenting the high technology with the best efficiencies, yields, and reasonable prices adapted to the Moroccan purchasing power. In particular, the main obstacle to its adoption is the high-security and weak support of African states. In economic and social terms, this project will create permanent jobs and achieve significant energy gains by reducing the energy bills costs for a range of sectors: housing sector, hotels, hospitals, schools and universities, sports sector (heated swimming water), industrial and agricultural sector. Strengthening the economic Moroccan–African industry through the manufacturing and marketing of solar water heaters and why not export. Finally, generalizing the use of these products would significantly reduce carbon dioxide emissions and combat the dangers of the gas heaters that kill dozens a year.

8. Conclusions

In this paper the thermosyphon solar water heater was presented, modeled, simulated and assessed according to six Moroccan climatic zones which represent the recent zoning carried out by the Moroccan Agency of Energy Efficiency (AMEE), in order to predict the system performances, and the dynamic behaviors of different parameters which are describing the thermosyphon. The system was modeled and simulations where launched under TRNSYS simulation program.

According to the simulation results the thermosyphon's performances in terms of the annual average solar fraction varies between 50% and 70%. The results confirmed that the climatic conditions are the key of the simulation. They have to be taken into consideration to optimize the overall system efficiency. Moreover, an economic study which is about the generalization of this technology was conducted to analyze the potential benefits offered by this solar solution in different zones in Morocco from a weather point of view, and to predict the total life-cycle savings. In fact, life-cycle savings were optimal in the first zone represented by Agadir, due to the optimal combination of performances and high needs that are concentrated in this zone. To complete this study, an environmental section was developed to exhibit the amount of CO_2 emissions that can be avoided by the implantation of this project. The results confirmed that a considerable amount of greenhouse gas emissions is reduced.

The penetration of large-scale solar water heaters in Morocco is prevented by many barriers, such as the understanding's lack among the population of the financial unattractiveness to poor households and the benefits they provide. However, thermosyphon will be accepted easier when the population understands all these technology advantages. In the case of SWH systems, the benefits are not immediately visible for a payback period of six years: the energy bill will be reduced, but not fast enough in order to justify the cost of installation. Thus, the governmental subsidies will be an important incentive solution for beneficiaries. The generalization of this type of installation will not be possible without the introduction of incentive mechanisms that involves the public authorities, subsidizing the purchase of solar water heaters which will enable citizens to equip themselves at a lower cost.

The advantages linked to the generalization of the thermosyphon in Morocco are multiple, both economically and environmentally. Citizens could save their energy bill in the long term, since after an average of six years they would have profitable purchase price. Moreover, manufacturing the solar water heater in Morocco will lower the import bill cost. Furthermore, it will also help the development of subcontracting and involve SMEs (Small and Medium Enterprises) in the manufacture of this product. Hence, entrepreneurship in this solar field will grow and be encouraged. Therefore, jobs will be created which will absorb unemployment.

Author Contributions: Conceptualization, F.Z.G. and A.A.; Methodology, F.Z.G. and A.A.; Software, F.Z.G. and A.A.; Supervision, T.K., A.J. and A.B.; Visualization, H.E.-H.; Writing—original draft, F.Z.G. and A.A. All authors have read and agreed to the published version of the manuscript.

Funding: SOL'R SHEMSY project is financed by IRESEN(Institut de Recherche en Energie Solaire et Energie Nouvelle).

Data Availability Statement: Data available in a publicly accessible repository.

Acknowledgments: The authors are grateful for the funding support provided by the Institut de Recherche en Energie Solaire et Energies Nouvelles (IRESEN) under the project SOL'R SHEMSY.

Conflicts of Interest: The authors declare no conflict of interest.

Nomenclature

A_c	Collector area (m^2).
F_R	Collector heat removal efficiency factor (-).
$F_R U_L$	Slope of the collector efficiency curve ($kJ\ h^{-1}\ K^{-1}\ m^{-2}$).
$F_{R(\tau\alpha)}$	Intercept efficiency corrected for non-normal incidence (-).
F	Friction factor (-).
F'	Collector efficiency factor (-).
U_L	Heat loss coefficient ($kJ\ h^{-1}\ K^{-1}\ m^{-2}$).
G_{test}	Collector mass flux at test conditions ($kg\ h^{-1}\ m^{-2}$).

G	Collector flow rate per unit area (kg h^{-1} m^{-2}).
ρ_i	Density of ith node (kg m^{-3}).
Δh_i	Height of the ith node (mm).
ΔP_i	Change in pressure across the ith node (bar).
h_{Li}	Frictional head loss in the ith node on the pipe (-).
H_p	The friction head losses on the pipes (-).
r	Ratio of collector heat removal efficiency factor, F_R, to the test conditions (-).
N_x	Number of equal sized collector nodes (-).
K	Correction coefficient for additional friction due to developing flow in the pipe (-).
k	Node number (-).
R_e	Reynolds coefficient (-).
f	Friction factor for flow in pipes (-).
v	Velocity of the fluid (m s^{-1}).
$\dot{m}$	Thermo siphon flow rate (kg h^{-1}).
C_p	Specific heat of working fluid (kJ kg^{-1} °C^{-1}).
g	Gravitational constant (N m^2 kg^{-2}).
$(\tau\alpha)$	Product of the cover transmittance and the absorber absorptance (-).
$(\tau\alpha)_b$	$(\tau\alpha)$ for beam radiation depends on the incidence angle (-).
$(\tau\alpha)_n$	$(\tau\alpha)$ at normal incidence (-).
$(\tau\alpha)_s$	$(\tau\alpha)$ for sky diffuse radiation (-).
$(\tau\alpha)_g$	$(\tau\alpha)$ for ground reflected radiation (-).
I_{bT}	Beam radiation incident of the solar collector (kJ h^{-1} m^{-2}).
I_d	Diffuse horizontal radiation (kJ h^{-1} m^{-2}).
I_T	Global radiation incident on the solar collector (kJ h^{-1} m^{-2}).
I_T	Total incident radiation per unit area (kJ h-1 m-2).
β	Collector slope above the horizontal plane (°).
θ	Incident angle for beam radiation (°).
b_0	Negative of the 2nd order coefficient in the IAM curve fit equation (-).
IAM	Incident angle modifier (-).
T_a	Ambient temperature (°C).
T_{ci}	Collector inlet temperature (°C).
T_{CO}	Collector outlet temperature (°C).
L_h	Length of collector headers (mm).
L_{co}	Length of collector outlet (mm).
L_{ci}	Length of collector inlet (mm).
d_i, d_o	Diameter of collector inlet and outlet pipes (mm).
L_i, L_0	Length on inlet and outlet piping (mm).
H_c	Vertical distance between outlet and inlet of collectors (mm).
H_O	Vertical distance between outlet of tank and inlet to collector (mm).
D_r	Riser tank diameter (mm).
D_h	Header tank diameter (mm).
V	Tank volume (L).
L	Pipe length (mm).
d	Pipe diameter (mm).
N_c	Number of parallel collector risers (-).
U_{api}	Conductance for heat losses from collector inlet pipe (kJ h^{-1} k^{-1}).
U_{Apo}	Conductance for heat losses from collector outlet pipe (kJ h^{-1} k^{-1}).
$(UA)_s$	Overall UA value for storage tank (kJ h^{-1} k^{-1}).
$Q_{aux,max}$	Maximum heating rate (kW).
P_{max}	Maximum pressure (bar).
K_T	Thermal losses of the tank ((kJ h^{-1} m^{-2}).
C_{sc}	Capacity of solar circuit (L).
E_{coll}	Efficiency of the collector (%).
F_{sol}	Solar fraction of the system (%).
Q_{aux}	Auxiliary heating load (kJ h^{-1}).
Q_{DHW}	Energy rate to load (kJ h^{-1}).
Q_{ucoll}	Energy rate from heat source (kJ h^{-1}).
I_{coll}	Total radiation on tilted surface (kJ h^{-1} m^{-2}).
ROI	Return on investment (years).

References

1. Williams, J.R. Solar energy: Technology and applications. *STIA* **1974**, *75*, 12425.
2. Ong, K. An improved computer program for the thermal performance of a solar water heater. *Sol. Energy* **1976**, *18*, 183–191. [CrossRef]
3. Bleu, P. Stratégie méditerranéenne pour le développement durable. In *Mise à Jour des Indicateurs de Suivi*; 2013.
4. Gergaud, O. *Modélisation Énergétique et Optimisation Économique d'un Système de Production Éolien et Photovoltaïque Couplé au Réseau et Associé à un Accumulateur*; 2002.
5. Anderson, R.; Kreith, F. Natural convection in active and passive solar thermal systems. In *Advances in Heat Transfer*; Elsevier: Amsterdam, The Netherlands, 1987; Volume 18, pp. 1–86.
6. Michaelides, I.; Lee, W.; Wilson, D.; Votsis, P. Computer simulation of the performance of a thermosyphon solar water-heater. *Appl. Energy* **1992**, *41*, 149–163. [CrossRef]
7. Carlsson, B.; Meir, M.; Rekstad, J.; Preiß, D.; Ramschak, T. Replacing traditional materials with polymeric materials in solar thermosiphon systems–Case study on pros and cons based on a total cost accounting approach. *Sol. Energy* **2016**, *125*, 294–306. [CrossRef]
8. Kalogirou, S.A.; Papamarcou, C. Modelling of a thermosyphon solar water heating system and simple model validation. *Renew. Energy* **2000**, *21*, 471–493. [CrossRef]
9. Tecchio, C.; Oliveira, J.; Paiva, K.; Mantelli, M.; Gandolfi, R.; Ribeiro, L. Thermal performance of thermosyphons in series connected by thermal plugs. *Exp. Therm. Fluid Sci.* **2017**, *88*, 409–422. [CrossRef]
10. Xu, Z.; Zhang, Y.; Li, B.; Wang, C.-C.; Ma, Q. Heat performances of a thermosyphon as affected by evaporator wettability and filling ratio. *Appl. Therm. Eng.* **2018**, *129*, 665–673. [CrossRef]
11. Vasiliev, L.L.; Grakovich, L.; Rabetsky, M.; Vassiliev, L.L., Jr.; Zhuravlyov, A. Thermosyphons with innovative technologies. *Appl. Therm. Eng.* **2017**, *111*, 1647–1654. [CrossRef]
12. Feliński, P.; Sekret, R. Effect of PCM application inside an evacuated tube collector on the thermal performance of a domestic hot water system. *Energy Build.* **2017**, *152*, 558–567. [CrossRef]
13. Kumar, S.; Dubey, A.; Tiwari, G. A solar still augmented with an evacuated tube collector in forced mode. *Desalination* **2014**, *347*, 15–24. [CrossRef]
14. Hassanien, R.H.E.; Li, M.; Tang, Y. The evacuated tube solar collector assisted heat pump for heating greenhouses. *Energy Build.* **2018**, *169*, 305–318. [CrossRef]
15. Essa, M.A.; Mostafa, N.H. Theoretical and experimental study for temperature distribution and flow profile in all water evacuated tube solar collector considering solar radiation boundary condition. *Sol. Energy* **2017**, *142*, 267–277. [CrossRef]
16. Koffi, P.; Andoh, H.; Gbaha, P.; Touré, S.; Ado, G. Theoretical and experimental study of solar water heater with internal exchanger using thermosiphon system. *Energy Convers. Manag.* **2008**, *49*, 2279–2290. [CrossRef]
17. Zerrouki, A.; Boumedien, A.; Said, N.; Tedjiza, B. Input/output test results and long-term performance prediction of a domestic thermosiphon solar water heater in Algiers, Algeria. *Renew. Energy* **2002**, *25*, 153–161. [CrossRef]
18. Tang, R.; Cheng, Y.; Wu, M.; Li, Z.; Yu, Y. Experimental and modeling studies on thermosiphon domestic solar water heaters with flat-plate collectors at clear nights. *Energy Convers. Manag.* **2010**, *51*, 2548–2556. [CrossRef]
19. Zeghib, I.; Chaker, A. Simulation of a solar domestic water heating system. *Energy Procedia* **2011**, *6*, 292–301. [CrossRef]
20. Shrivastava, R.; Kumar, V.; Untawale, S. Modeling and simulation of solar water heater: A TRNSYS perspective. *Renew. Sustain. Energy Rev.* **2017**, *67*, 126–143. [CrossRef]
21. Jordan, U.; Vajen, K. Influence of the DHW load profile on the fractional energy savings: A case study of A solar combi-system with TRNSYS simulations. *Sol. Energy* **2001**, *69*, 197–208. [CrossRef]
22. Tanha, K.; Fung, A.S.; Kumar, R. Performance of two domestic solar water heaters with drain water heat recovery units: Simulation and experimental investigation. *Appl. Therm. Eng.* **2015**, *90*, 444–459. [CrossRef]
23. Hobbi, A.; Siddiqui, K. Optimal design of a forced circulation solar water heating system for a residential unit in cold climate using TRNSYS. *Sol. Energy* **2009**, *83*, 700–714. [CrossRef]
24. Ge, J.; Wu, J.; Chen, S.; Wu, J. Energy efficiency optimization strategies for university research buildings with hot summer and cold winter climate of China based on the adaptive thermal comfort. *J. Build. Eng.* **2018**, *18*, 321–330. [CrossRef]
25. Bellia, L.; Borrelli, M.; De Masi, R.F.; Ruggiero, S.; Vanoli, G.P. University building: Energy diagnosis and refurbishment design with cost-optimal approach. Discussion about the effect of numerical modelling assumptions. *J. Build. Eng.* **2018**, *18*, 1–18. [CrossRef]
26. Chen, S.; Yang, J. Loop thermosyphon performance study for solar cells cooling. *Energy Convers. Manag.* **2016**, *121*, 297–304. [CrossRef]
27. Solomon, A.B.; Daniel, V.A.; Ramachandran, K.; Pillai, B.; Singh, R.R.; Sharifpur, M.; Meyer, J. Performance enhancement of a two-phase closed thermosiphon with a thin porous copper coating. *Int. Commun. Heat Mass Transf.* **2017**, *82*, 9–19. [CrossRef]
28. Kousksou, T.; Bruel, P.; Cherreau, G.; Leoussoff, V.; El Rhafiki, T. PCM storage for solar DHW: From an unfulfilled promise to a real benefit. *Sol. Energy* **2011**, *85*, 2033–2040. [CrossRef]
29. Dwaikat, L.N.; Ali, K.N. The economic benefits of a green building–Evidence from Malaysia. *J. Build. Eng.* **2018**, *18*, 448–453. [CrossRef]

30. Rezvani, S.; Bahri, P.; Urmee, T.; Baverstock, G.; Moore, A. Techno-economic and reliability assessment of solar water heaters in Australia based on Monte Carlo analysis. *Renew. Energy* **2017**, *105*, 774–785. [CrossRef]
31. Sokka, L.; Sinkko, T.; Holma, A.; Manninen, K.; Pasanen, K.; Rantala, M.; Leskinen, P. Environmental impacts of the national renewable energy targets–A case study from Finland. *Renew. Sustain. Energy Rev.* **2016**, *59*, 1599–1610. [CrossRef]
32. Bouhal, T.; Agrouaz, Y.; Allouhi, A.; Kousksou, T.; Jamil, A.; El Rhafiki, T.; Zeraouli, Y. Impact of load profile and collector technology on the fractional savings of solar domestic water heaters under various climatic conditions. *Int. J. Hydrog. Energy* **2017**, *42*, 13245–13258. [CrossRef]
33. Allouhi, A.; Jamil, A.; Kousksou, T.; El Rhafiki, T.; Mourad, Y.; Zeraouli, Y. Solar domestic heating water systems in Morocco: An energy analysis. *Energy Convers. Manag.* **2015**, *92*, 105–113. [CrossRef]
34. Arouri, M.E.H.; Youssef, A.B.; M'henni, H.; Rault, C. Energy consumption, economic growth and CO_2 emissions in Middle East and North African countries. *Energy Policy* **2012**, *45*, 342–349. [CrossRef]
35. Kousksou, T.; Allouhi, A.; Belattar, M.; Jamil, A.; El Rhafiki, T.; Arid, A.; Zeraouli, Y. Renewable energy potential and national policy directions for sustainable development in Morocco. *Renew. Sustain. Energy Rev.* **2015**, *47*, 46–57. [CrossRef]
36. Ministry of Energy, Water and Environment, Department of the Environment. *Oasis in Morocco: Adaptation to Climate Change. Resilient Oasis: Adaptation to Climate Change*; 2008.
37. Goldemberg, J.; Coelho, S.T. Renewable energy—Traditional biomass vs. modern biomass. *Energy Policy* **2004**, *32*, 711–714. [CrossRef]
38. Morocco HPCI. Available online: https://www.hpci-office.jp/cgi-bin/hpcidatabase/app/summary.cgi?view=list&col=f70&date_from=&date_to=&doclang=all&invisiblecol=0&outerhpci=1&prjcls=field&resrc=65551&row=C&s=hp160268&select=32&sop=and&lang=en (accessed on 16 December 2020).
39. El-Katiri, L. *OCP Policy Center*; 2016.
40. Gonçalves, H. Solar Heating and Cooling in Portugal. In *Solar Heating and Cooling Programme-International Energy Agency, Energy Day*; 2018.
41. REN 21. *Renewables 2019–Global Status Report*; 2019.
42. Bennouna, A. *Trois Décennies D'énergie Solaire Thermique au Maroc*; EcoActu: Casablanca, Morocco, 2019.
43. Bennouna, A. *Trois Décennies D'énergie Solaire Photovoltaïque au Maroc*; EcoActu: Casablanca, Morocco, 2019.
44. Moroccan Agency of Energy Efficiency A.M.E.E. Available online: http://www.amee.ma (accessed on 23 December 2020).
45. TRNSYS Studio Documentation, Mathematical. Available online: https://sel.me.wisc.edu/trnsys/ (accessed on 16 December 2020).

Article

Energy Efficiency for Social Buildings in Morocco, Comparative (2E) Study: Active VS. Passive Solutions Via TRNsys

Fatima Zohra Gargab [1,2], Amine Allouhi [1], Tarik Kousksou [2], Haytham El-Houari [1,3,*], Abdelmajid Jamil [1] and Ali Benbassou [1]

1   Ecole Supérieure de Technologie de Fes, Université Sidi Mohamed Ben Abdellah, Route d'Imouzzer, Fes BP 2427, Morocco; fatimazohra.gargab@gmail.com (F.Z.G.); allouhiamine@gmail.com (A.A.); Abdelmajid.jamil@usmba.ac.ma (A.J.); Ali.benbassou@usmba.ac.ma (A.B.)
2   Laboratoire des Sciences de l'Ingénieur Appliquées à la Mécanique et au Génie Electrique (SIAME), Université de Pau et des Pays de l'Adour/E2S UPPA, EA4581, 64000 Pau, France; tarik.kousksou@univ-pau.fr
3   Laboratoire de Mathématiques, Modélisation en Physique Appliquée, Ecole Normale Supérieure de Fès, U.S.M.B.A, Route Bensouda, Fes BP 5206, Morocco
*   Correspondence: haythamelhouari1@gmail.com

Abstract: This paper aims to highlight the potential of solar water heater installations in Morocco. The project involves the comparison of active and passive solutions for energy efficiency in buildings. To this end, a numerical simulation model of solar water heater installations is created under TRNsys. Three hot water demand scenarios (Low, Standard, and High) were taken into account for the six climatic zones defined in the Moroccan thermal regulation of constructions. The same software (TRNsys) is used to model a pilot building consisting of 16 flats. Energy efficiency actions have been applied to the building envelope (insulation and glazing) and simulations are made for the six areas. The simulation results comparing energy and financial savings show the influence of subsidized gas prices on solar water heaters' relevance despite significant energy savings. This work proves that solar water heaters will be a primary obligation for Morocco, taking into account changes in butane gas prices.

Keywords: solar system; domestic hot water production; solar water heaters; individual and collective solar water heater systems; dynamic simulation; TRNbuild; TRNSYSstudio

Citation: Gargab, F.Z.; Allouhi, A.; Kousksou, T.; El-Houari, H.; Jamil, A.; Benbassou, A. Energy Efficiency for Social Buildings in Morocco, Comparative (2E) Study: Active VS. Passive Solutions Via TRNsys. *Inventions* **2021**, *6*, 4. https://doi.org/ 10.3390/inventions6010004

Received: 22 October 2020
Accepted: 23 December 2020
Published: 28 December 2020

Publisher's Note: MDPI stays neutral with regard to jurisdictional claims in published maps and institutional affiliations.

1. Introduction

The Moroccan energy situation assessment has revealed that the country imports about 95% of its energy needs, 60% of which is oil, equivalent to 9 million TOE (tons of oil equivalents) [1]. This national energy bill weighs heavily where 20% of total imports and 50% of the trade deficit represent only oil. Despite this situation, Morocco supports gas users, industry, and households where domestic gas price support subsidies are close to 10% of investment expenditure in the general state budget [2].

In the same directive of this evaluation, and focusing on the residential sector that represents the purpose of this work, it is essential to discuss the country's electric consumption. Electrical energy consumption increased from 487.38 kWh/person in 2000 to 911.64 kWh/person in 2014. Meanwhile, projections of energy needs for the 2020s expect a growth of 5% to 6% for electricity consumption (which has recorded much more values especially for households due to COVID-19 quarantines and curfews), which should be around 1000 kWh per inhabitant annually [3]. Household consumption surveys from 1985 to 2001, despite limited human development, have shown that spending on electricity and butane increased by 9.1% per year [4]. Besides, the average consumption of ONE's residential customers is close to 75 kWh per month, less than 1 kWh/m^2 covered [5]. At the same time, the need for air conditioning (heating and cooling) is not negligible in many of the Kingdom's major cities.

Therefore, this country's energy dependence, its policy, and the energy bill imposed by the residential sector present real constraints that weigh on the current chances of the country's sustainable development. To cope with this situation, Morocco is committed to an energy efficiency policy, expecting to achieve a primary energy saving of 15% by 2030 by establishing an energy efficiency plan in all sectors (residential, tertiary, industry and agriculture) [6]. Among these sectors, the building is the second largest consumer of energy with a 25% share of the country's total energy consumption, of which 18% is reserved for residential and the rest for the tertiary sector [7]. This consumption is distributed on all of the building needs, of which domestic hot water represents a very important part, and one of the most exciting applications of solar energy in various configurations, the individual solar water heaters (SWH) and collective solar installations (CSI) of residential and tertiary demand characterized by large and regular hot water needs throughout the year. In Morocco, currently most of these needs are satisfied using electric and gas water heaters, despite the high petroleum product tariffs introduced in February 2006 and the current electric kWh prices which make the SWH an appreciated solution than all conventional sources, except for zero-rated or subsidized fuel (butane) scenarios [8,9].

This paper involves comparing active and passive energy efficiency solutions for building. The results presented come from the application of first the insulation, double glazing, and finally, all conditions required by Moroccan thermal regulation of building construction as passive solutions. Solar water heaters for the production of domestic hot water are an active solution. The numerical simulation via TRNsys makes it possible to compare different scenarios of the studied solutions, developing a complete comparison. In the literature, several studies have taken the same direction, but a limited number have treated the comparison and the combination of the two action types; they have developed each solution separately. The global aim is to enhance building energy efficiency. Cabeza et al. [10] have tested the active and passive system on an experimental building prototype located in Puigverd de Lleida, Spain. His study aims to evaluate the energy savings resulting from testing, first different technologies of active solutions such as solar thermal, free cooling, geothermal, etc. Secondly, testing sustainable materials and phase change materials as passive solutions to designing the green building envelope. Gou et al. [11] combined the two design strategies (active and passive) to develop the first zero energy building in Southeast Asia. This work is based on an existing building and under tropical climate conditions, presenting a cost-effectiveness comparison of passive and active strategies. The results show that passive solutions must be applied on a large scale to have significant economies because of its long payback period compared to the active ones. Another work presented by Buonomano et al. [12] converges in the same direction. The passive and active effects on the photovoltaic and thermal system's integration were analyzed by evaluating performance, energy demands, and electrical production. The numerical model was validated and compared to an experimental model integrated into an office building simulated in several European climates; the results achieved a 56.8% to 104.4% energy consumption reduction. All works take this direction intending to improve energy efficiency [13] in different building types and models, the newest and the existing [14] even for the historical ones [15], under several conditions of climate, occupation, construction, and system integration to achieve optimal thermal comfort.

The renewable energy solutions integrated into the building are massively present in the literature. Yan Wang et al. [16] developed a new energy performance index by studying the integration of renewable energy on school buildings for Germany's heating needs. The numerical model was developed and simulated; a new performance indicator was developed to evaluate the active solution proposed; the borehole outlet temperature effects and heat recovery efficiency. The results show that the distribution enhanced can achieve a proper thermal comfort level for the case studied, reduce electricity costs, and reduce CO_2 emissions by 5.3 kg per typical winter.

Bougiatioti et al. [17] studied the integration of active solar systems on an existing building in traditional settlements to develop several possibilities of a smart architectural integration in Cyprus and Greece, where the construction conditions make this action advantageous. This work aims to evaluate gains and difficulties highlighting the need to exploit these benefits under the economic crisis.

In the same directive of highlighting the active solutions role in both energy performance and the indoor built environment, Amilios et al. [18] studied the active solar systems integration on southern Europe building envelopes. The studies have been carried out under the inhabitants' comforts requirements, making an energy and environment assessment. This action gives a remarkable reduction of cooling and heating loads in the proposed comprehensive environmental approach. In the present paper, the active solution proposed is the solar water heater (SWH) under two configurations; the individual and the collective. There are numerous studies in the literature focused on the performance and design optimization of residential water heaters, independent of passive solutions, as a performing solution [19,20]

Building efficiency studies using passive solutions are pronounced in the literature. Whether the optimization is the key of several works, Chen et al. [21] optimized a typical passive residential building design in China. The optimal model was carried out under the influence of five weather conditions investigating a sensitivity analysis related to ventilation, outdoor thermal, and solar radiation, studying the proposed model's applicability. The passive model was studied in several cases in several locations and climate conditions such as hot and humid climates [22].

In the present work, there are three types of passive solutions proposed. The first solution is improving the building envelope by the insulation, where several works introduced many innovative insulation materials for optimal building efficiency [23]. Kaushik Biswas et al. [24] integrate composite insulation boards containing foam-encapsulated vacuum insulation panels. The material's thermal characterization was done developing a new process manufacturing technology called modified atmosphere insulation to release a significant cost reduction keeping a high performance. Others incorporate the anchors into the insulation in order to strengthen the insulation [25]. The dynamic material was also present as an insulation proposition [26], and many other innovative materials were proposed as an excellent cheap alternative such as bio insulation [27] and wood waste [28].

The insulation studies focused on materials [29] and other parameters such as thickness [30], which was analyzed by Cemek et al. [31]; they develop a parametric study based on insulation thickness, taking into account several indicators such as energy savings, payback period, and CO_2 emission reductions under the climate conditions of different Turkish cities.

The second passive solution proposed is glazing, which was studied broadly. In the literature, several works detail this solution, focusing on different materials and developed glazing types such as aerogel glazing, PDLC (Polymer Dispersed Liquid Crystal) switchable glazing, and simple and double types [32–36].

The present comparison presented was projected on a social building, taking into account different solutions for each type. The study integrated the technological side of active solutions, presenting a comparison between flat plate collectors and the evacuated tube one under two configurations; the individual and the collective, comparing the solar solution with the conventional ones presented in the market, and studying all existing combinations, in order to prove the importance of investing in those solutions compared to passive propositions. This work will defend a suitable solution between a passive and active one for the Moroccan context by taking into account all dimensions such as economic, energetic and the payback period, with the best solution needing to combine all those parameters for an optimal result.

2. Methodology and Case Study

The building studied was modeled with the TRNsys Studio 16 software and TRN-BUILD module. The final model makes the energy demand profile evaluation possible, taking into account critical building parameters such as wall composition, facade orientation, zone volume, occupation, and set point temperature. We studied the envelope improvement by using the separate passive solutions of insulation and glazing, integrating complete Moroccan thermal regulations, and applying the active system (solar water heaters), to compare all propositions.

In this section, the building model is presented under its architecture, with the composition of walls and construction materials of the first proposition which will be improved. All simulation hypotheses and parameters will be exhibited (occupancy, ventilation, zoning and masks). Then the solutions studied will be described under inputs and characteristics and the simulation model developed for each proposition (building envelope improvement, individual SWH, and collective SWH).

All simulations were carried out under Moroccan climatic conditions. The Moroccan territory has been subdivided into six climatic zones that are homogenous and circumscribed: Zone 1, Zone 2, Zone 3, Zone 4, Zone 5, and Zone 6. This climatic zoning was adopted for the thermal regulation of buildings in Morocco, and the simulations were done for the six climatic Moroccan zones presented in Table 1.

Table 1. Meteorological data.

Climatic Zone	Climate	Temperature $^\circ$C
Zone 1	Subtropical-semiarid	14.3–23.2
Zone 2	Mediterraneanhot	12.6–24.2
Zone 3	Mediterranean/continental	9.6–26
Zone 4	Humidtemperateclimate	6.2–25.4
Zone 5	Semiarid	12.4–28
Zone 6	Hot desert	9.1–33.2

2.1. The Building Models

2.1.1. Building Architecture

The building studied in this project is social housing with 300 m^2; it consists of four levels: a ground floor and three floors. The accommodation is composed of 16 apartments, and the main facade faces west. Figure 1 shows the global modeling of the building with the Google Sketch UP software.

Figure 1. The building model with the Google Sketch UP software.

Each floor of the building has four standard apartments of 70 m^2, except the apartments on the ground floor, which are 60 m^2 in size. Figure 2 shows the architectural plan of the ground floor and a representative floor.

Figure 2. The architectural plan of the ground floor and a representative floor.

2.1.2. Building Envelope (Construction Materials)

The composition and construction materials of the building are shown in Table 2.

Table 2. The conventional building envelope.

Walls	Layers	Thickness (mm)	U (W/m²·K)
External wall	External plaster	20	0.847
	Brick	200	
	Plasterboard	30	
Internal wall	Plasterboard	25	1.183
	Brick	50	
	Plasterboard	25	
Low floor	Heavy concrete	200	2.949
	Mortar	50	
	Floor tile	10	
High floor	Plasterboard	13	2.683
	Slab	120	
	Lightweight concrete	200	
	Mortar	50	
	Floor tile	10	
Glazing	Simple glazing	-	5.8

2.1.3. Simulation Building Model and Hypothesis

Figure 3 shows the simulation model flowchart on TRNsys studio software. The building model was created using TRNBUILD connected to TRNsys by the construction model (type 56). It is a non-geometric scale model with one air node per zone, representing the air volume's zone heat capacity and the capacities that are closely related to the air node (furniture, for example). Thus, the node's capacity is a separate entry in addition to the volume of the zone.

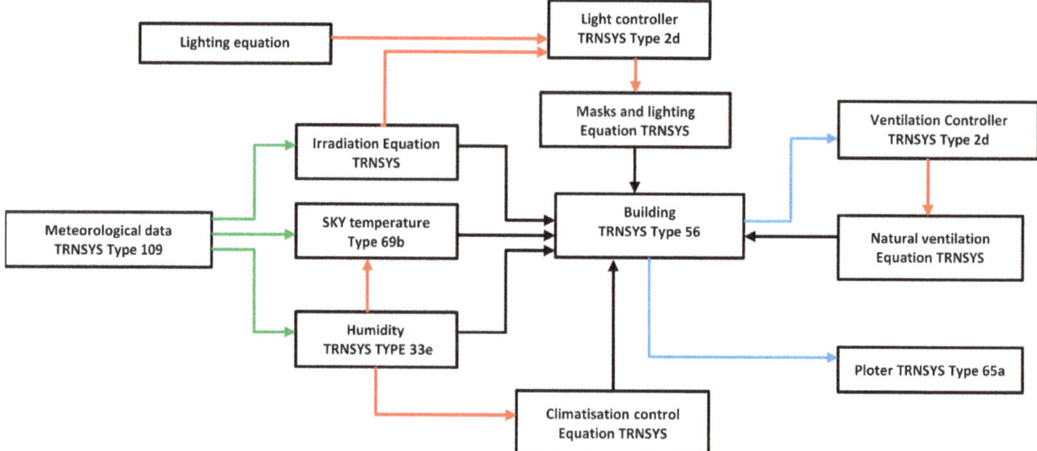

Figure 3. TRNsys information diagram for multi-zone building.

The thermal zones: the selected model was considered with 36 thermal zones, 16 heated and air-conditioned thermal zones. Each zone includes the living rooms and rooms of each apartment and 20 thermal buffer zones, which are the bathrooms, kitchens, and halls of all the apartments. Figure 4 represents the thermal zones considered. Two other hypotheses would have been possible: to model each room by a thermal zone and model each floor as a single zone.

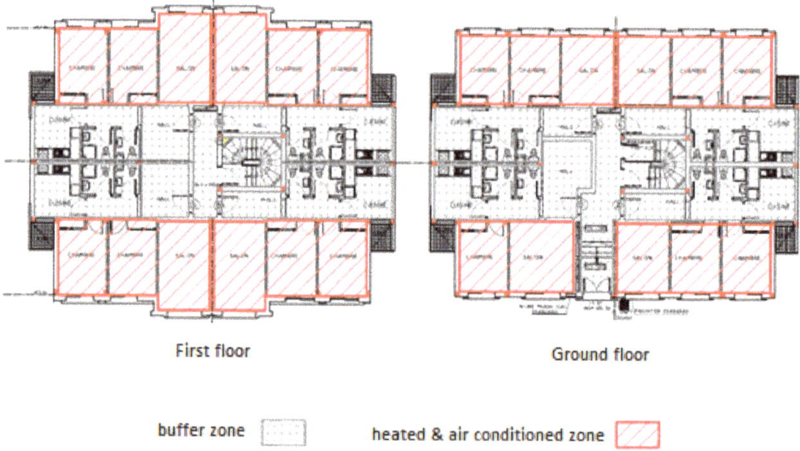

Figure 4. The building zoning.

2.1.4. Simulation Hypotheses

It is impossible to reproduce an entire building numerically, and the choice of modeling assumptions is essential. The main assumptions that have been made for the modeling of the building in question are:

Masks: the balconies and the progress of the roof were considered. Thermal solar evacuated tube collectors are composed of vacuum tubes allowing part of the radiation to penetrate, and flat plate thermal solar collectors create shading. However, it was assumed that this shading was minimal and was not considered. No close mask was considered (trees, buildings).

Internal inputs: TRNsys offers the possibility to specify the people's thermal load according to ISO 7730 and VDI 2078. The VDI 2078 is a German standard. That is why we used the international standard ISO 7730. This standard specifies people's heat according to their activity types. Each apartment is supposed to be occupied by four seated people releasing 100 watts according to the ISO 7730 standard. The lighting heat input is about 10 W/m^2. Besides, the heat of electrical appliances constitutes additional inputs.

Ventilation: The imposed ventilation rates depend on many aspects and differ according to the regulations. Table 3 groups the values imposed by the RT2012 standard. Since the building in question is a social building, the ventilation mode chosen during the simulation is natural. In heated and air-conditioned areas, the air exchange rate is equal to 0.6 volume/hour. For areas not treated by air conditioning and heating-kitchen, bathroom, and shower, the rate of air renewal chosen is equal to 2.6 volume/hour.

Table 3. Regulatory ventilation flow rates.

Number of Principal Rooms Per Housing	Extracted Flow Rate (m^3/h)				
	Kitchen	Principal Bathroom	Secondary Bathrooms	Wash Room	
				Unique	Multiple
1	20/75	15	15	15	15
2	30/90	15	15	15	15
3	45/105	30	15	15	15
4	45/120	30	15	30	15
5 and more	45/135	30	15	30	15

Heating and air conditioning: For heating and cooling, the heating set temperature is equal to 20 °C, and that of the air conditioning is equal to 26 °C. The systems are active at all times under a constant schedule profile and the value of the control is 1.

2.1.5. The Passive Solutions

Four dynamic thermal simulation scenarios were made to evaluate heating and cooling demands. The first conventional scenario concerns the case without any energy efficiency action on the building envelope; it is the study reference case previously presented. The three others are carried out using insulation and double glazing separately, then all requirements of the Moroccan building thermal regulations. The simulation hypotheses (internal inputs, masks, thermal zoning, heating setpoint, air conditioning, etc.) are the same for the four scenarios.

2.1.6. Insulation

The purpose of insulation during this scenario is to respect the thermal transfer coefficient's values specified by the thermal regulation of Moroccan buildings. The insulation type and thickness are not discussed in this work. The building wall thermal characteristics and floors in different Moroccan thermal zones are shown in Table 4. The windows are single glazing types with a heat transfer coefficient of U = 5.8 $W/m^2 \cdot K$.

Table 4. Thermal characteristics of insulated building walls and floors according to the Moroccan standards.

Thermal Zones	U (W/m²·K)			
	External Walls	Internal Walls	Low Floor	High Floor
Zone 1	0.847	1.813	2.949	0.750
Zone 2	0.847	1.813	2.949	0.636
Zone 3	0.847	1.813	2.949	0.657
Zone 4	0.60	1.813	2.949	0.552
Zone 5	0.60	1.813	2.949	0.636
Zone 6	0.60	1.813	2.949	0.636

2.1.7. Glazing

In this scenario, the only action to be taken on the building envelope is to respect the heat transfer coefficient values set by the Moroccan thermal regulation of construction (MTRC) for glazing. The building will not be isolated in the six zones' first simulation scenario, using the single glazing. Table 5 lists the thermal characteristics of the building envelope for the second scenario.

Table 5. Building envelope action on the glazing.

Walls	Layers	Thickness (mm)	U (W/m²·K)
External wall	Exterior plaster	20	0.847
	Brick	200	
	Plaster panel	30	
Internal wall	Plaster panel	25	1.183
	Brick	50	
	Plaster panel	25	
Floor	Heavy concrete	200	2.949
	Mortar	50	
	Tile	10	
Roof	Plaster panel	13	2.683
	Slab	120	
	Light concrete	200	
	Mortar	50	
	Tile	10	
Glazing	Double glazing	-	2.83

2.1.8. Moroccan Thermal Regulation of Construction (MTRC)

In this scenario, the MTRC requirements are met for all zones. Table 6 illustrates the thermal characteristics of the envelope adopted for the simulations.

Table 6. Thermal characteristics of the building envelope under Moroccan thermal regulation of construction (MTRC) requirements.

Envelope	U (W/m²·K)					
	Zone 1	Zone 2	Zone 3	Zone 4	Zone 5	Zone 6
External wall	0.847	0.847	0.847	0.60	0.60	0.60
Internal wall			1.813			
Floor			2.949			
Roof	0.750	0.636	0.636	0.552	0.636	0.636
Glazing			2.83			

2.1.9. The Active Solutions

The load profile is adopted for the scenarios studied (individual and collective) as the building schedule need for domestic hot water (Figure 5), where the hourly coefficient represents the hot water demand at 50 °C in 10 min.

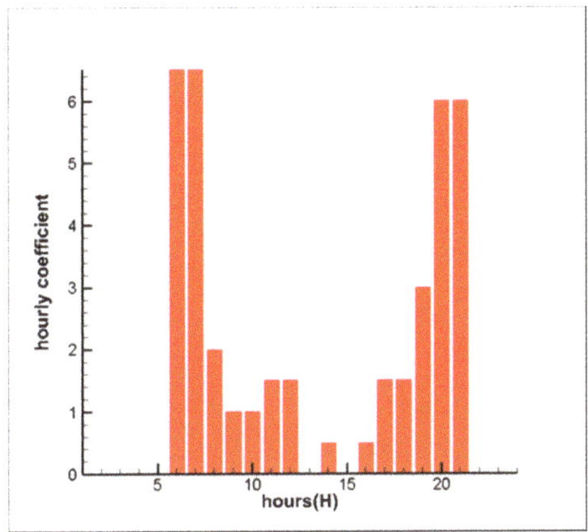

Figure 5. The daily profile of domestic hot water use.

Taking into account the habits of Moroccan consumers of the average social class, the daily need per person for domestic hot water is set at 25 L/day/person for low demand. 37.5 L/day/person for average demand. and 50 L/day/person for a high demand, also by estimating an occupancy of 4 people per apartment.

2.1.10. Individual SWH System

For the individual simulation of solar system (Figure 6), two types of solar water heaters were considered: the glazed flat plate collector and the evacuated tube collector. Both models of the individual solar system are realized under TRNsys. These used the existing library models: a glazed flat plate collector (FPC) and evacuated tube collector (ETC), the load profile model, a storage tank model, and cold-water temperature variation model. Table 7 shows the technical characteristics of the collectors and tanks used in this study scenario.

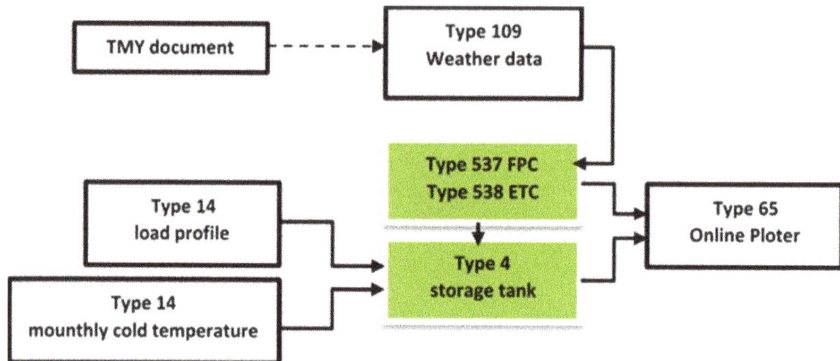

Figure 6. The flowchart of individual solar water heater (SWH) system.

Table 7. Technical characteristic of collectors and tanks (individual SWH).

Collectors		
Technical Characteristics	**FPC**	**ETC**
Area (m^2)	1.8–3.3	1.5–3
Tested flow rate kg/hr·m^2	60	60
Optical coefficient (%)	0.7	0.7
Thermal loss coefficient a1 (kJ/hr·m^2 K)	3.13	1.894
Thermal loss coefficient a2 (kJ/hr·m^2 K^2)	0.016	0.0039
Mass flow rate (min; max) (kg/hr)	0.90	0.90
Optimal inclination (°)	45	45
Tank		
Volume (L)	150–200–300	
Specific heat of the fluid (kJ/kg·K)	4.19	
Fluid density (kg/m$^{2)}$	1000	
Thermal loss coefficient (kJ/hr·m^2·K)	1.4	
Height (m)	1	
Nodes number	3	
Auxiliary heater power (kW)	1–1.5–3	
Maximum auxiliary heater temperature (°C)	60	

2.1.11. Collective SWH System

The apartment layout is symmetrical. Eight apartments on each side of the building have a common technical shaft. The collective solar system has been treated in two parts, with eight apartments per part that make the entire building (16 apartments). The results obtained will be identical. The configuration of the simulation's solar system model is the collective solar water heater with collective storage and individualized back-up, which is divided into two circuits: the primary circuit or the solar circuit, which includes the field of collectors, and the solar tank with a capacity that covers the eight apartments' needs; this circuit is located in the roof relating to the solar collectors and the principal tank. The second circuit contains the individualized tanks with an integrated back-up; it relates to the principal tank and the individual tanks for each apartment, a circulation pump, and controllers which manage the pump activation/deactivation referring to set up temperatures. Figure 7 illustrates the connection between the primary and secondary circuits of the collective solar water heater system. Table 8 shows the technical characteristics of the solar tank used in this scenario. For collectors, the same technology types used on the individual case are adopted.

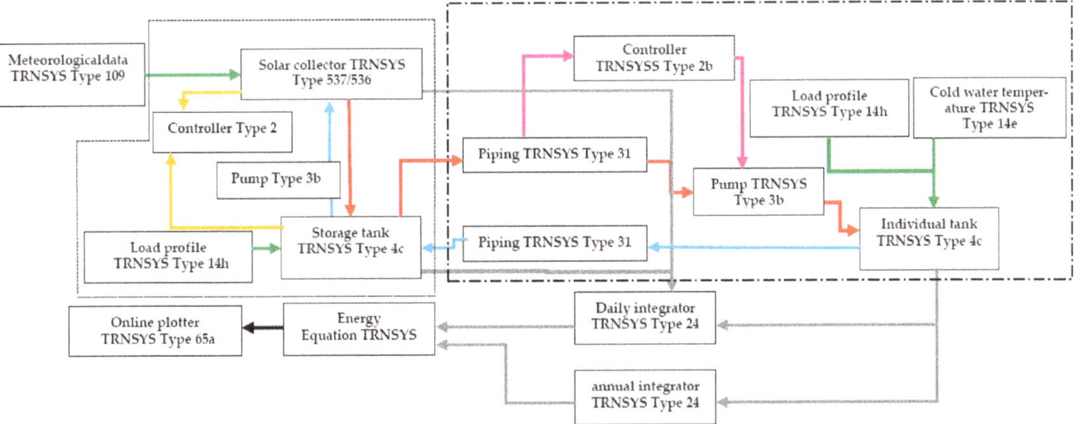

Figure 7. Simulation information flow diagram of the collective solar system.

Table 8. Technical characteristic of collectors and tanks (collective SWH).

	Tank	
	Common	**Individual Auxiliary Tank**
Volume (L)	1600	200
Specific heat of the fluid (kJ/kg·k)	4.19	
Fluid density (kg/m^2)	1000	
Thermal loss coefficient (kJ/hr·m^2·K)	1.4	
Height (m)	1.8	1
Nodes number	6	3
Auxiliary heater power (kW)	0	1.5
Maximum auxiliary heater temperature (°C)	60	

3. Results and Discussion

3.1. Energy and Efficiencies

Passive Solutions

Table 9 shows the values obtained following simulation of the building modeled in TRNBUILD, for one year with a time step of one hour. The conventional building's thermal simulation results without any energy efficiency actions clearly show the significant energy demand for heating and cooling, especially for zones 6 and 4 which exceed 200 kWh/m^2/year. A decrease between 23% and 39% of total annual needs is achieved with insulation compared to the conventional case. This ratio remains higher than that set by the MTRC. The ratio decrease is vital for the heating demand reaching a 44% decline for zone 5 (Marrakech), against the annual requirement ratio in air conditioning that does not significantly decline. The decline is zero for zone 1 represented by Casablanca city; the maximum recorded is a 27% decrease for zone 6 (Errachidia).

Table 9. Heating and cooling annual demand.

Zone	Zone 1				Zone 2				Zone 3				Zone 4				Zone 5				Zone 6			
	C	I	G	TR	C	I	G	TR	C	I	G	TR	C	I	G	TR	C	I	G	TR	C	I	G	TR
Heating annual demand (kWh/m²/year)	41	25	35	15	56	32	48	14	95	55	81	29	160	92	128	47	48	27	38	13	81	46	65	23
Cooling annual demand (kWh/m²/year)	30	29	31	27	40	37	43	34	58	44	56	39	40	30	39	30	93	71	87	61	132	96	121	66
Total annual demand(kWh/m²/year)	71	54	66	42	96	69	91	48	153	99	137	68	200	122	167	77	141	98	125	74	213	142	186	89

C: the conventional building (reference case without improvement actions); **I**: the conventional building with insulation integration; **G**: the conventional building applying the glazing action; **TR**: the building adapted to the Moroccan thermal regulation of construction (MTRC).

The conformity of the glazing's thermal characteristics with the (MTRC) makes it possible to reach a maximum decrease of 25% on the total annual ratio compared to the conventional case. This value is recorded in zone 5 (Marrakech). The ratio of air conditioning needs is not significantly reduced. This drop does not exceed 8% for zone 6 (Errachidia).

After having carried out the building dynamic thermal simulations for different thermal zones, and under the fourth scenario condition (MTRC) the insulation and glazing, the sum of the heating and cooling power ratios for each zone is greater than the one set by the MTRC; this is justified by the fact that the building is misdirected, which increases solar gain.

On the other hand, we compared this to the results obtained by the simulations of the conventional, insulated, and glazed scenarios. In zone 4 represented by Ifrane, a maximum saving of 62% is achieved between the conventional case and the MTRC scenario. This remark is made for the other three scenarios where maximum savings are achieved in zone 4.

It is concluded that the application of a double-glazing solution in buildings, without the application of insulation, does not lead to significant savings in comparison with the cost of investment—a comparative economic study will be presented afterwards. On the other hand, insulation alone makes it possible to reach a savings threshold that is considered reasonable.

From this, the application of the MTRC allows significant energy consumption savings of buildings even if, in this case study, the results were higher than the values set by the MTRC. This difference is explained by the orientation of the building, which is oriented towards the west. We can also add that the simulation does not recreate the reality without error, given the large number of assumptions made in this study.

3.2. Active Solutions (SWH)

3.2.1. Individual Solar Water Heater

Figure 8a,b, Figure 9a,b, and Figure 10a,b show the monthly variation of useful energy extracted from the tank, the extra energy and solar fraction for the three different demands (low, standard, and high), for the two technology types. The collected energy varies according to demand; this is due to catchment surface variation. The collected energy varied between 214 kWh and 146 kWh in August and February, respectively, for the low demand case (150 L) and between 311 kWh and 214 kWh for the standard demand case (200 L). For high demand (300 L), energy varied between 400 kWh in August and 274 kWh in February.

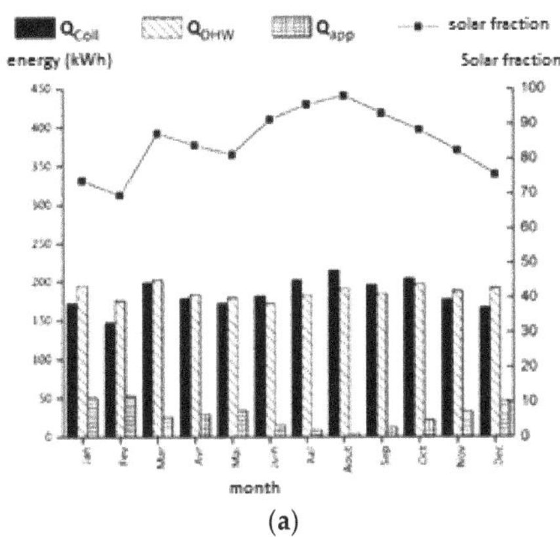

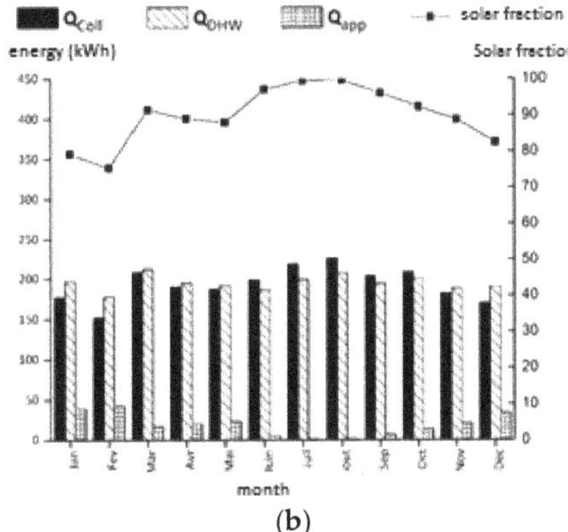

Figure 8. (**a**) Flat plate collector, (**b**) evacuated tube collector (low demand).

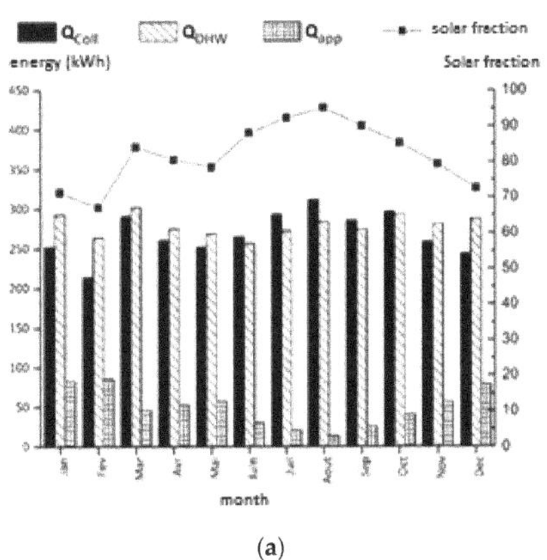

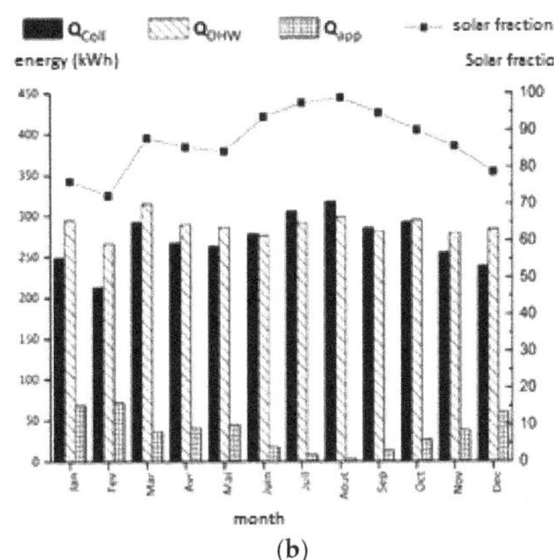

Figure 9. (**a**) Flat plate collector, (**b**) evacuated tube collector (standard demand).

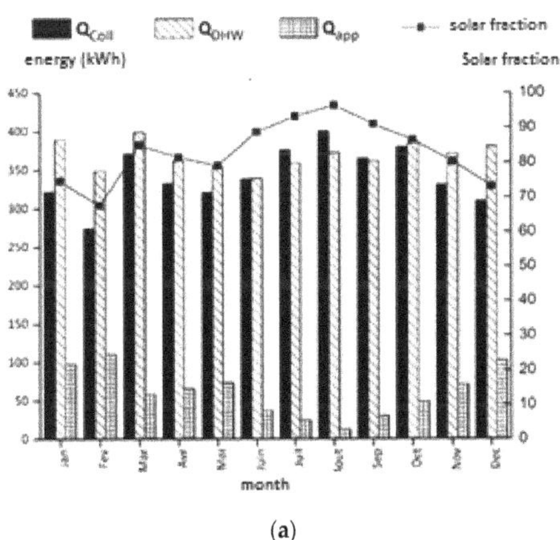

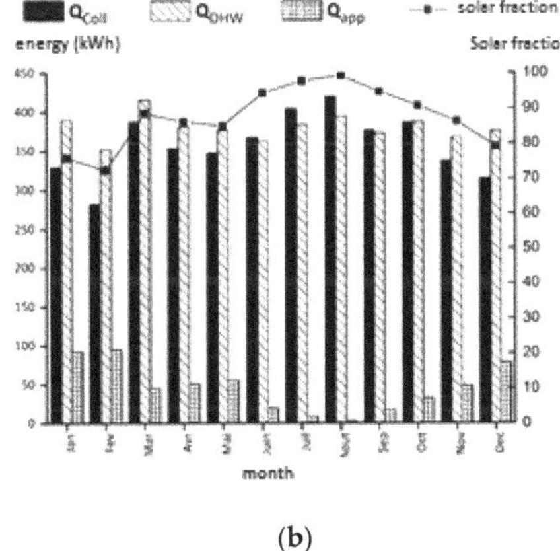

Figure 10. (**a**) Flat plate collector, (**b**) evacuated tube collector (high demand).

The energy extracted from the solar storage tank varied according to the load profile. The solar fraction varies in the same way for the three cases. It varied between 70% in February and 98% in August.

The energy collected by the solar thermal collectors (evacuated tube collectors (ETC), and flat plate collectors (FPC)) varies in function of the loads' profiles; this is due to the load profile influence on the collector's area. The surface increases with the increase in the amount of water exhausted.

The energy extracted from the tank, which means the energy required to heat the sanitary water, varies during the year, depending on the cold-water temperatures. This energy is higher than the useful energy produced during the year's cold period, and the auxiliary system compensates for this difference. Note that the backup system is requested in the summer. This can be explained by the demand for hot water during the night and the morning's first hours.

The solar fraction varies at the same rate for the three types of use. This fraction reaches a maximum value in the summer by the order of 99%.

The two solar water heater technologies comparison, based on the results presented in this section, makes it possible to conclude that the evacuated tube collectors perform better than the glazed flat plate collectors in the third zone. To analyze the performance of the two technologies in the Moroccan context and draw comparative conclusions, simulations of solar water heaters are made for all thermal zones of Morocco. The analysis of the simulation results shows that evacuated tube solar water heaters are better than glazed flat plate collectors, during cold seasons of the year, in areas with a Mediterranean climate in the North and Oceanic in the West. On the other hand, the two technologies are identical for the warm seasons. For the fourth zone, we note that the vacuum tubes are more efficient from March. In areas that have a continental climate zone 5 and a Saharan climate zone 6, the vacuum tube collectors have a better performance than flat glass collectors throughout the year.

3.2.2. Collective Solar Water Heater

Simulations of solar systems for domestic hot water production make it possible to evaluate collectors and production systems' performance. Figures 11a and 12a summarize the simulation results for flat plate and vacuum tube collectors. It can be seen that the useful energy produced by the vacuum tube collectors is higher than that resulting from the use of glazing flat plate collectors during the cold months. This is explained by the round shape of the tubes, which enables them to collect different radiation types (direct, reflected, etc.) even on cloudy days, unlike plans that have a lower yield.

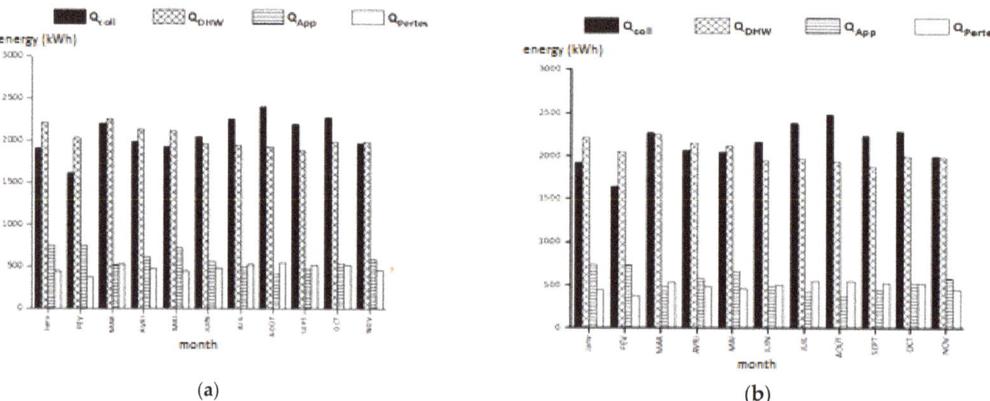

Figure 11. (**a**) Flat plate collector, (**b**) evacuated tube collector.

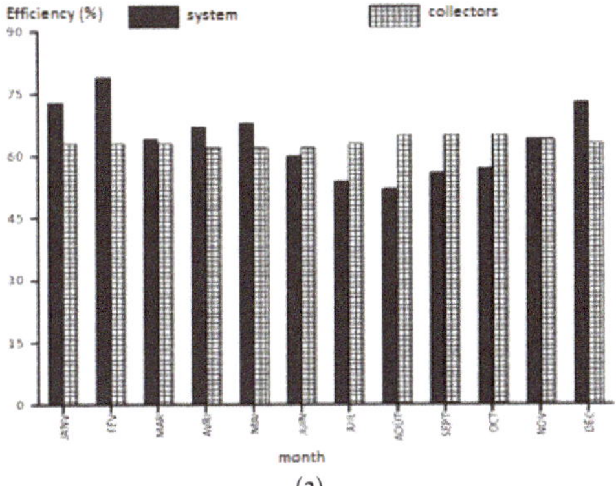
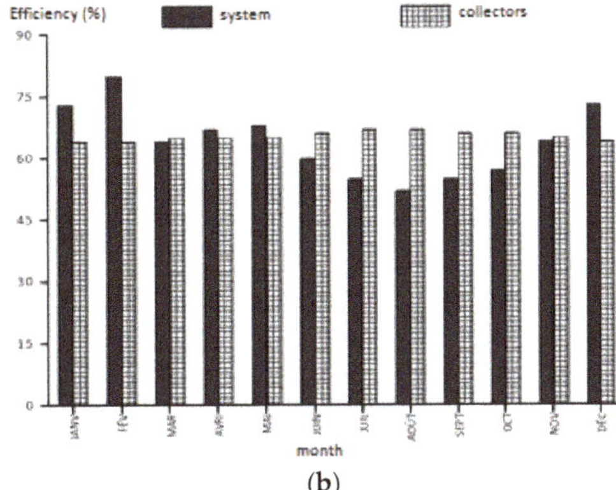

Figure 12. (**a**) Flat plate collector, (**b**) evacuated tube collector (efficiencies).

Along the day, thermal losses produced in pipes and storage tanks are almost identical throughout the year, causing the auxiliary heater's operation (backup system) all year, with a decrease during hot months.

The efficiency of the domestic hot water production system based on the simulations carried out shows that the vacuum tube system's performances are higher than the flat plate collector. During the cold months, the efficiency of the vacuum tube system reaches 80% in February. However, the efficiency of both systems is identical during the summer months. The vacuum tube collectors' field efficiency is always greater than that of the FPC, as shown in Figures 11b and 12b.

The results of collective solar system simulations related to domestic hot water production show that the solar water heaters with ETC are more efficient than FPC, whatever the demand (low, standard, or high) under the same load profile.

An analysis of the solar fractions and energies provided concludes that the evacuated tubes are more efficient than the glazed flat plate technology on all of Morocco's cold zones (zones 4, 5, and 6). For zones 1 and 2, characterized by their moderate climates, we note that the two technologies are identical the year except for the winter months, where ETC systems deliver more energy than the FPC collectors.

3.3. Economies

3.3.1. Hot Water

The hot water production is ensured in the Moroccan context in general by instantaneous gas water heaters in the first place, or by instantaneous electric water heaters. There are also storage water heaters. Table 10 summarizes the overall yields of the various domestic hot water systems considered in this study.

Table 10. Efficiency of hot water production systems.

System		Global Efficiency
Electrical water heater	Night accumulation	0.7
	Instantaneous	0.95
Gas accumulation water heater		0.5
Instantaneous gas water heater	With pilot light	0.6
	Without pilot light	0.8

The energy comparison focused only on the individual SWH systems. The comparison of the collective case requires a study of the conventional collective systems of hot water production.

The power required to produce domestic hot water varies according to the climate zone which system thermal loss variation and specific cold-water temperature for each location (variation of cold-water temperature during the year). Figures 13–15 represent the annual energy demand variation for domestic hot water production for the various conventional systems in the different Moroccan thermal zones according to the demand for domestic hot water.

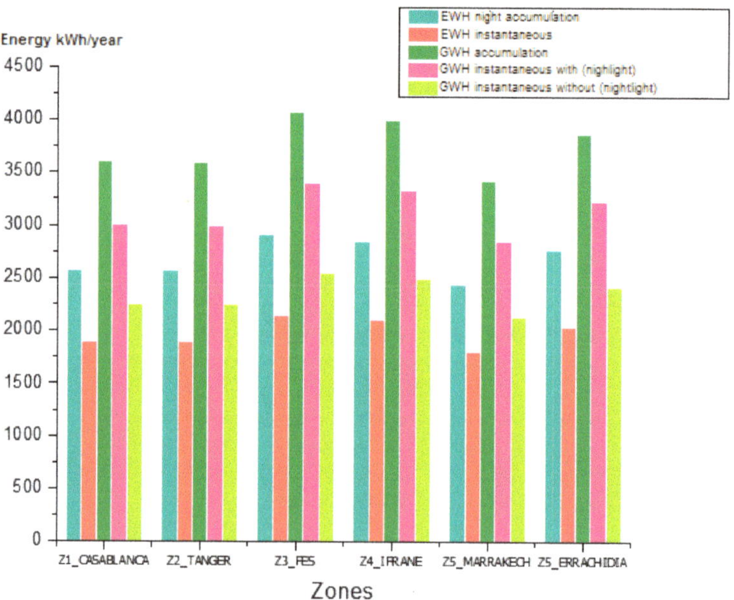

Figure 13. SWH production energy for low demand.

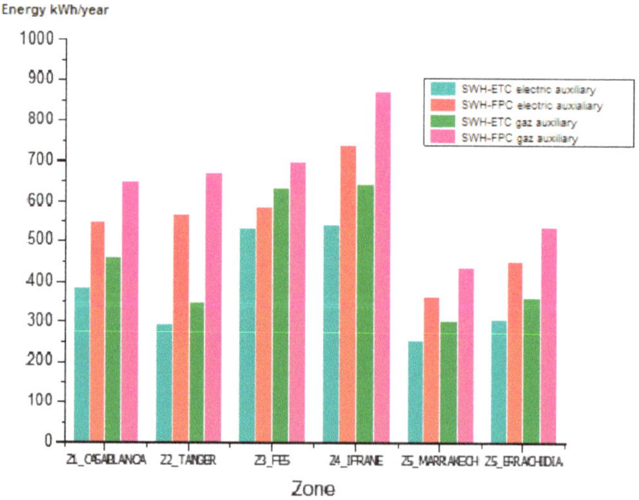

Figure 14. Annual energy of the backup system for standard demand of hot water.

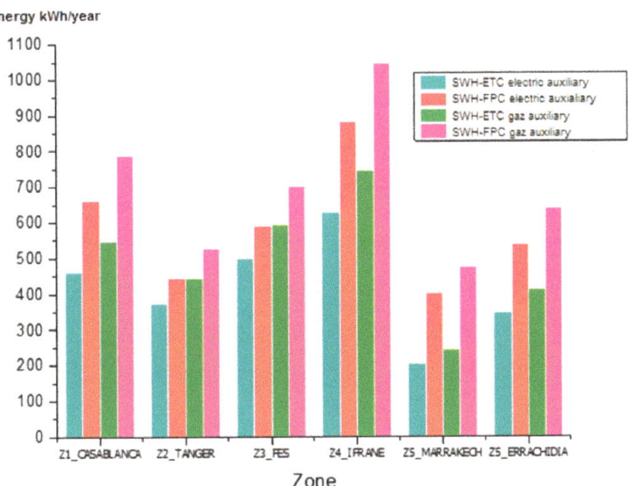

Figure 15. Annual energy of the backup system for high demand of hot water.

It was found that the gas storage water heater is the most energy-consuming of all systems, followed by the gas instantaneous water heater. The instantaneous electric water heater is the most economical technology. The required annual energy thresholds vary according to the thermal zones; this is due to the variation of each zone's cold-water temperatures.

This part concerned the savings evaluation of domestic hot water energy production and consumption by conventional systems and solar production systems. We can see a maximum energy saving of 97% by replacing a gas water heater in zone 5 (Marrakech) by a solar water heater ETC system with an electric auxiliary heater. The minimum savings obtained was about 58%, replacing an instantaneous electric water heater in zone 4 (Ifrane) by a solar water heater with FPC technology with gas back-up.

3.3.2. Passive Solutions

The total energy demand for the building is represented per m^2 heated per year. Dynamic thermal simulations have evaluated three solutions:

- Building with insulation respecting the MTRC requirements.
- The glass building respecting the MTRC requirements.
- Building respecting all MTRC requirements.

The three solutions are compared with the conventional case. Figure 16 shows the ratios obtained for each thermal zone.

The MRTC allows a significant reduction of demand for heating required by the conventional case (reference case). It is found that the glazing solution (double glazing for most areas) does not give a significant saving compared to the insulation solution. It is shown on Table 11.

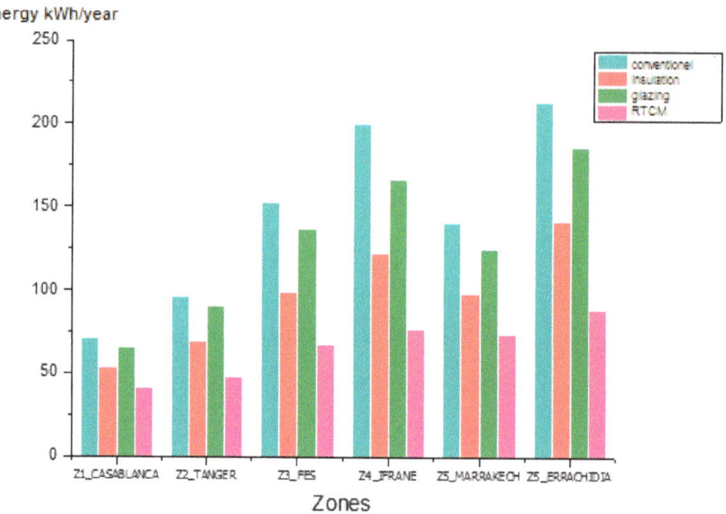

Figure 16. Air conditioning and heating consumption ratios.

Table 11. Savings of passive solutions in kWh/year.

Zones	Insulation	Glazing	MTRC
Zone 1	680	200	1160
Zone 2	1080	200	1920
Zone 3	2160	640	3400
Zone 4	3120	1320	4920
Zone 5	1720	640	2680
Zone 6	2840	1080	4960

This work aims to compare the savings generated by active and passive energy efficiency actions. Table 12 groups the energy-saving quantities using the solar water heater system compared to the conventional water system.

Tables 11 and 12 show that the solar water heater solution saves up to 7649 kWh/year, and in the worst case, it is possible to reach 1227 kWh/year, which is higher than 44% of the savings generated by passive solutions. Financial analysis is presented to complete this energy analysis, considering the savings and investments for different actions.

Table 12. Energy savings achieved for the different scenarios of SWH production.

	A→1	A→2	A→3	A→4	B→1	B→2	B→3	B→4	C→1	C→2	C→3	C→4	D→1	D→2	D→3	D→4	E→1	E→2	E→3	E→4
								Energy Savings Achieved for Different Scenarios of DHW Production (kWh/year)												
Z1	2328	2204	2282	2134	1650	1526	1604	1456	3359	3234	3312	3164	2758	2633	2711	2563	2006	1882	1959	1812
Z2	2319	2193	2271	2122	1642	1516	1595	1445	3347	3221	3300	3150	2747	2621	2700	2550	1997	1871	1950	1800
Z3	2639	2578	2588	2516	1873	1812	1823	1750	3803	3742	3752	3680	3124	3063	3073	3001	2275	2214	2225	2152
Z4	2498	2116	2431	1978	1747	1365	1680	1227	3639	3257	3572	3119	2973	2591	2906	2453	2141	1759	2074	1621
Z5	2284	2168	2254	2116	1641	1524	1610	1472	3262	3146	3232	3094	2692	2575	2661	2523	1978	1862	1948	1810
Z6	2573	2461	2537	2404	1845	1733	1808	1675	3681	3568	3644	3511	3035	2923	2998	2865	2227	2115	2191	2058
Z1	3477	3316	3404	3213	2460	2299	2387	2196	5023	4862	4950	4759	4121	3960	4049	3857	2994	2833	2921	2730
Z2	3561	3292	3506	3186	2546	2277	2491	2171	5104	4834	5049	4729	4204	3935	4149	3829	3079	2810	3024	2704
Z3	3830	3776	3730	3665	2681	2627	2581	2517	5576	5522	5476	5411	4558	4503	4457	4393	3284	3230	3184	3120
Z4	3737	3542	3636	3404	2611	2416	2510	2278	5449	5254	5348	5116	4451	4256	4349	4117	3203	3008	3101	2869
Z5	3412	3299	3364	3230	2447	2334	2399	2265	4879	4766	4831	4697	4023	3910	3975	3841	2954	2840	2906	2771
Z6	3845	3699	3788	3614	2753	2606	2695	2521	5506	5360	5449	5275	4537	4391	4480	4306	3326	3180	3269	3095
Z1	4690	4489	4603	4364	3334	3133	3247	3008	6751	6550	6665	6425	5549	5348	5462	5223	4046	3845	3959	3720
Z2	4769	4699	4699	4616	3415	3346	3345	3263	6826	6756	6756	6673	5626	5556	5556	5473	4126	4056	4056	3973
Z3	5321	5229	5228	5118	3790	3698	3696	3587	7649	7557	7556	7446	6291	6199	6198	6088	4594	4502	4500	4391
Z4	5076	4825	4958	4660	3575	3323	3457	3158	7359	7107	7241	6942	6027	5776	5909	5611	4363	4112	4245	3946
Z5	4685	4489	4646	4414	3398	3202	3359	3127	6641	6445	6602	6370	5500	5304	5461	5229	4073	3878	4035	3803
Z6	5189	4998	5124	4897	3733	3541	3668	3441	7404	7213	7339	7112	6112	5921	6047	5820	4497	4306	4433	4205

A: electric DWH night storage, B: Instant electric DWH, C: Gas water heater, D: Instant gas water heater with pilot light, E: Instant gas water heater without pilot light; 1: Evacuated tube SW with electric auxiliary heater, 2: Flat plate SWH with electric auxiliary heater, 3: Evacuated tube SW with gas auxiliary heater, 4: Flat plate SWH with electric auxiliary heater.

Low hot water demand. Standard hot water demand. Hight hot water demand.

3.4. Financial Analysis

It is not enough to determine the electricity bill based on the difference between the conventional case energy demand and after applying energy efficiency actions to evaluate the savings achieved. The incremental billing system influences the results.

The electricity bill savings require evaluating the conventional case bill and the new one after applying the energy efficiency actions.

Table 13 summarizes the annual energy bill for conventional domestic hot water systems, as presented previously in this paper.

Table 13. Annual invoice for hot water conventional systems production.

SWH	Zones	Amount MAD/Year							
		A	**B**	**C1**	**C2**	**D1**	**D2**	**E1**	**E2**
LOW HOT WATER DEMAND	Z1	2553	1818	949	2966	791	2472	593	1854
	Z2	2547	1814	947	2960	789	2467	592	1850
	Z3	2939	2082	1072	3350	893	2792	670	2094
	Z4	2873	2037	1051	3284	875	2737	657	2053
	Z5	2408	1714	900	2815	750	2345	563	1759
	Z6	2773	1969	1019	3186	849	2655	637	1992
STANDARD HOT WATER DEMAND	Z1	4087	2867	1423	4449	1186	3708	890	2781
	Z2	4077	2860	1420	4440	1184	3700	888	2775
	Z3	4773	3297	1607	5025	1339	4187	1005	3140
	Z4	4655	3223	1576	4926	1313	4105	985	3079
	Z5	3840	2699	1351	4222	1125	3518	844	2639
	Z6	4479	3114	1529	4780	1274	3983	956	2987
HIGHT HOT WATER DEMAND	Z1	5862	3998	1898	5932	1581	4943	1186	3708
	Z2	5848	3989	1894	5920	1578	4933	1184	3700
	Z3	6798	4668	2143	6700	1786	5583	1339	4187
	Z4	6636	4551	2101	6568	1751	5474	1313	4105
	Z5	5499	3760	1801	5629	1501	4691	1125	3518
	Z6	6391	4379	2039	6373	1699	5311	1274	3983

Table 14 shows the annual energy bill for the auxiliary system consumptions of the solar water heaters, based on the simulation results.

Table 14. Energy bill of auxiliary systems.

SWH	Zones	Amount MAD/Year					
		ETC_A	**FPC_A**	**ETC_E1**	**FPC_E1**	**ETC_E2**	**FPC_E2**
LOW HOT WATER DEMAND	Z1	224	336	78	116	243	364
	Z2	228	341	79	118	247	370
	Z3	244	299	85	104	264	324
	Z4	320	664	111	230	347	720
	Z5	145	250	50	87	157	271
	Z6	175	276	61	96	190	300
STANDARD HOT WATER DEMAND	Z1	349	495	121	172	379	536
	Z2	266	509	92	177	289	552
	Z3	482	531	167	184	522	575
	Z4	488	664	169	230	529	720
	Z5	230	332	80	115	249	360
	Z6	276	408	96	142	299	442
HIGHT HOT WATER DEMAND	Z1	417	600	145	208	452	649
	Z2	337	399	117	138	365	433
	Z3	449	532	156	185	487	577
	Z4	566	803	197	275	615	860
	Z5	185	361	64	125	200	391
	Z6	312	484	108	168	338	525

Table 15 summarizes the maximum and minimum savings achieved after replacing the conventional system (gas water heater) by the solar one. The next step in this comparative study is calculating the annual savings achieved by passive energy efficiency solutions. It is assumed that the heating and cooling systems are electric, so it is necessary to initially calculate the energy bills of the various scenarios presented previously, which were devoted to the building's simulations. It is to deduce the maximum and minimum values of the savings generated, comparing it with the solar water heater savings results. The saving was calculated by taking into account two scenarios based on Moroccan government gas subsidies.

Table 15. Amount saved by installing a solar water heater for an individual building.

SWH	Zones	Amont MAD/Year			
		With Subsidies		Without Subsidies	
		Max	Min	Max	Min
LOW HOT WATER DEMAND	Z1	2888	229	2475	257
	Z2	2881	222	2468	251
	Z3	3265	346	2854	371
	Z4	3173	155	2762	211
	Z5	2765	292	2358	313
	Z6	3125	337	2712	361
STANDARD HOT WATER DEMAND	Z1	4328	354	3966	395
	Z2	4348	336	3985	379
	Z3	4858	430	4606	474
	Z4	4757	265	4486	321
	Z5	4142	484	3760	512
	Z6	4684	514	4383	548
HIGHT HOT WATER DEMAND	Z1	5787	537	5717	586
	Z2	5803	751	5731	785
	Z3	6642	762	6642	807
	Z4	6439	453	6439	510
	Z5	5565	734	5435	764
	Z6	6283	749	6283	790

The financial amounts presented in this part are represented in MAD (Moroccan dirham) which is equivalent on 14 December 2020 to 0.11 USD.

In Morocco, the state subsidies varied following the fall in the price of butane gas, with its average unit subsidy increasing from 4840 MAD/Ton in 2018, or 58 DH per 12 kg cylinder and 14.5 MAD per 3 kg cylinder, to 3652 MAD/Ton in 2019 (January–September period), corresponding to 44 MAD per 12 kg cylinder and 11 MAD per 3 kg cylinder. In the present work, the economic study was carried out referring to 2019 prices.

3.5. Building

Tables 16 and 17 show the heating and cooling energy bill—the same energy reductions generated by energy efficiency actions reflected the energy bills. Glazing does not realize significant gains. However, insulation or both solutions combined are more significant. The glazing solution does not exceed 18% of savings compared to the insulation. On the other hand, the insulation reaches 44%, the combination of the two solutions, which represent the MTRC, reaches a reduction of 66% compared to the conventional case.

Table 16. Annual energy bill for heating and cooling.

Zones	Energetic Bills MAD			
	Conventional	Insulation	Glazing	MTCR
Z1	808	575	737	445
Z2	1131	756	1030	552
Z3	1870	1118	1640	735
Z4	2388	1342	1952	821
Z5	1640	1079	1440	799
Z6	2824	1753	2439	1088

Table 17. Savings rate achieved.

Zones	Percentage of Savings		
	Insulation	Glazing	MTCR
Z1	29%	9%	45%
Z2	33%	9%	51%
Z3	40%	12%	61%
Z4	44%	18%	66%
Z5	34%	12%	51%
Z6	38%	14%	61%

The generated savings comparison of the solar water heaters production and the passive energy efficiency actions result in the maximum solar water heaters savings being more significant than passive actions. The most unfortunate cases of solar water heaters savings are close to those reached by passive actions, sometimes even lower.

3.6. Economies Comparative Ratios; Active vs. Passive

In this part, the rate represented by the economy reached by each solution proposed during this work is defined. Given the large number of possibilities of conventional domestic hot water systems and the replacing action by solar water heaters, the following cases are treated:

3.6.1. Passive Configurations

Configuration 1: electric heating and cooling, electric SWH.
Configuration 2: electric heating and cooling, gas SWH.

3.6.2. Active Configurations

A: Instantaneous electric water heater.
B: Instantaneous gas water heater without pilot light.
1: SWH-ETC with electric auxiliary heater.
2: SWH-ETC with gas auxiliary heater.
3: SWH-FPC with electric auxiliary heater.
4: SWH-FPC with gas auxiliary heater.
All results are for a standard domestic hot water application (volume of 200 L).

According to Tables 18 and 19, the solar water heater for configuration 1, where the domestic hot water production system is electric, achieves savings rates on the building energy bill higher than that achieved by applying passive actions in all Moroccan thermal zones.

Table 18. Savings using the active system SWH.

	Savings Ratio Configuration 1								Savings Ratio Configuration 2							
	Subsidies															
	Yes	No	Yes	No	Yes	No	Yes	No	Yes	No	Yes	No	Yes	No	Yes	No
	A→1		A→2		A→3		A→4		B→1		B→2		B→3		B→4	
Z1	69%	69%	75%	68%	65%	65%	65%	65%	32%	68%	45%	67%	23%	64%	42%	63%
Z2	65%	65%	69%	64%	59%	59%	59%	59%	31%	64%	39%	64%	19%	58%	35%	57%
Z3	54%	54%	61%	54%	54%	54%	54%	54%	18%	53%	29%	52%	16%	52%	29%	51%
Z4	49%	49%	54%	48%	46%	46%	46%	46%	15%	47%	24%	47%	10%	44%	22%	43%
Z5	57%	57%	60%	56%	55%	55%	55%	55%	25%	56%	31%	56%	21%	54%	29%	53%
Z6	48%	48%	51%	47%	46%	46%	46%	46%	18%	47%	23%	46%	14%	44%	22%	44%

Table 19. Savings using the passive system SWH (on total energy building).

	Savings Ratio					
Zones	Insulation		Glazing		MTRC	
	Config 1	Config 2	Config 1	Config 2	Config 1	Config 2
Z1	6.34%	13.72%	1.93%	4.18%	9.88%	21.38%
Z2	9.40%	18.57%	2.53%	5.00%	14.51%	28.68%
Z3	14.55%	26.16%	4.45%	8.00%	21.97%	39.48%
Z4	18.64%	31.01%	7.77%	12.93%	27.93%	46.46%
Z5	12.93%	22.58%	4.61%	8.05%	19.38%	33.86%
Z6	18.04%	28.33%	6.48%	10.19%	29.24%	45.93%

Under the second configuration conditions, the active solution is more interesting for all thermal zones than insulation and glazing actions. However, the MTRC saves more than the solar water heaters in the second configurations in the hottest areas.

The solar water heater solution with FPC in the case of configuration 1, where the domestic hot water production system is electric, allows saving rates on the building energy bill, almost identical to the second configuration in case of subsidies. In contrast, the SWH-ETC, compared to the conventional solution, generates considerable savings without subsidies.

In both configurations, it can be seen that solar water heater solutions deliver significant savings on buildings' energy bills compared to passive energy efficiency solutions.

3.7. Payback Time

Another critical factor in helping to make investment decisions in energy efficiency actions is the time to return on investment. The actions investments and the time of return on investment are evaluated.

3.7.1. Active System

The study treats two cases; individual and collective. The amounts given in results, including the prices of the various installation components and accessories, the maintenance costs, and the defective parts' replacement estimates, are not considered.

It can be seen in Table 20 that the payback time for a solar water heater project in the current case of subsidized butane gas prices is not attractive; the payback time in this case is more than 18 years which makes the project excessively uninteresting. On the other hand, for electric systems and unsubsidized butane gas, the payback times are almost identical depending on the demand; with more demand the return on investment (ROI) is lower, and the project becomes interesting.

Table 20. Return on investment (ROI) time of individual solar water heaters.

SWH	Zones	ROI (Year)					
		ETC			FPC		
		A	E1	E2	A	E1	E2
Low demand	Z1	8.6	37.3	8.4	6.9	39.9	6.8
	Z2	8.7	37.8	8.5	7.0	40.8	6.8
	Z3	7.5	32.3	7.4	5.7	27.6	5.7
	Z4	8.0	40.8	7.9	7.5	36.4	7.4
	Z5	8.8	32.9	8.5	7.0	32.7	6.8
	Z6	7.7	29.8	7.6	6.1	28.4	6.0
Standard demand	Z1	6.3	29.1	6.5	5.4	32.3	5.6
	Z2	6.1	25.3	6.3	5.4	33.6	5.6
	Z3	5.6	30.1	5.9	4.6	26.9	4.9
	Z4	5.8	31.7	6.1	5.0	39.7	5.3
	Z5	6.4	25.7	6.5	5.4	24.9	5.5
	Z6	5.5	23.2	5.8	4.7	23.3	4.9
High demand	Z1	4.8	22.4	5.2	4.2	24.3	4.6
	Z2	4.7	20.4	5.1	4.0	18.2	4.3
	Z3	4.1	19.4	4.6	3.4	17.7	3.9
	Z4	4.3	23.1	4.9	3.8	27.9	4.3
	Z5	4.8	18.4	5.2	4.2	18.7	4.5
	Z6	4.2	17.9	4.7	3.7	18.0	4.1

ROI's collective installations are calculated according to ETC and FPC systems and under the conditions of a standard demand for domestic hot water. In a collective building, the demand for hot water is not identical. Investments for this case were calculated without including maintenance charge prices and replacement estimates for defective parts. The assessment system consists of:

- 12 solar collectors,
- 2 solar tanks of 1600 L,
- 16 tanks of 200 L
- 400 m of piping,
- 16 electronic controllers,
- 16 circulation pumps,
- 2 copper tube collectors.

Following the same methodology of the individual part, the time of return on investment was calculated under two scenarios:

- Replace a conventional electric system with the solar water heater.
- Replace a conventional gas system with the solar water heater.

A is the electric solar water heater, E1: Gas is an instantaneous water heater without a pilot light and with subsidies. E2 is a gas instantaneous water heater without a pilot without subsidies.

Table 21 shows that the payback time for subsidized butane gas prices is relatively high, making collective solar projects unattractive in the current economic context. On the other hand, in the case of an electric system or unsubsidized butane gas, the return on investment does not exceed 8 years in most Moroccan zones. Moreover, in the zone of Fes, the time of return on investment on a collective solar project using FPC technology does not exceed 7.5 years.

Table 21. Return on investment time of collective solar water heaters.

Zones	ROI (Year)					
	ETC			FPC		
	A	E1	E2	A	E1	E2
Z1	8.7	40.4	9.0	8.6	51.6	8.9
Z2	8.4	35.2	8.7	8.7	53.8	9.0
Z3	7.8	41.8	8.2	7.4	43.0	7.8
Z4	8.0	44.0	8.4	8.0	63.5	8.4
Z5	8.9	35.6	9.1	8.6	39.8	8.8
Z6	7.7	32.2	8.1	7.5	37.2	7.9

3.7.2. Passive Actions

In the technical study, the insulation type was not very important because its purpose is to compare the savings of energy efficiency actions and not insulation specific technical studies. For this, the economic study data prices represent the average of the current prices on the market. Table 21 summarizes the insulation prices and the calculated return on investment time. It is noted that the time of return on investment is relatively short for insulation projects, and not more than a year except for zone 5, where it reaches two years. Furthermore, the double-glazing solution for zones 2 and 3 is not attractive, as the investment payback time exceeds 25 years. On the other hand, in zones 4 and 6 the double-glazing solution is fascinating; the ROI varies between 4 and 5 years, which is considered reasonable. This solution's payback time in zones 3 and 5 is 8 years, which makes this solution moderately interesting.

The application of the MTRC consists of combining the two previous solutions (insulation and glazing improvements).

The high payback time of the glazing improvement projects shown in Table 22 is reduced by the combination of various building improvements according to the MTRC. It is found that the ROI for zones 1 and 2, which was 25 years for glazing and one year for insulation, is reduced to 5 years for zone 1 and 3 years for zone 2 after applying the MTRC. Investment in the MTRC scenario is considered attractive, given the reduced ROI observed for all Moroccan climatic zones. The results obtained show that, despite the energy savings achieved with collective solar hot water production systems, low butane gas costs in Morocco and initial installation costs are detrimental to the project's profitability.

Table 22. Payback time of passive actions.

Zones	Investment MAD			Payback Time (year)		
	Insulation	Glazing	MTRC	Insulation	Glazing	MTRC
Z1	13,209	81,000	94,209	1.2	25.3	5.1
Z2	17,850	81,000	98,850	1	25.3	3.2
Z3	23,205	81,000	104,205	0.7	7.9	1.9
Z4	53,360	81,000	134,360	1.1	3.8	1.7
Z5	48,005	81,000	129,005	1.7	7.9	3
Z6	48,005	81,000	129,005	1.1	4.7	1.6

Passive energy efficiency solutions deliver significant savings in zones with a cold winter and a hot summer season.

In summary, it is difficult to make a clear and definitive judgment on comparing the savings made by solar solutions to produce domestic hot water and the actions on the enveloped building, but the active solutions are still more attractive than the passive ones.

4. Conclusions

Four building simulation scenarios under TRNsys were adopted (conventional, insulated, glazing improvements, and applying the Moroccan standard requirements). The glazing solution does not exceed 6% in savings on total energy demand (including the electrical demand for domestic hot water), while insulation reaches 17% in savings in zone 4. MTRC is the best solution. This solution allows for saving 27% of energy consumption.

The study's active solutions were under various domestic hot water production scenarios using solar systems, introducing different types of solar collectors, and the variables of the backup systems.

The energy demand comparison of the different conventional domestic hot water production systems and the new actions showed that solar solutions are remarkably efficient. In the same context, ETC and FPC comparisons show that vacuum tube collectors perform better in all Moroccan thermal zones during the winter. During the summer season, both technologies have solar fraction thresholds close to 90%.

The study of the collective solar water heating system on TRNsys reinforces the findings made in individual scenarios: the vacuum tube sensor is significantly efficient.

The savings made by applying the various solutions was two-part: energy and financial. A comparison of energy savings shows that solar water heater solutions are superior to those achieved by building envelope actions. The financial analysis of the savings on the calculated bills shows that solar water heater solutions are attractive in non-subsidized butane gas prices.

This comparative study opens the possibilities of integrating collective solar water heaters in buildings, considering the possible savings. From another point of view of this study's results, we push to seek the financing solutions of the collective solar projects possible, by green funds proposed by some non-governmental organizations or state organizations.

Author Contributions: Conceptualization, F.Z.G. and A.A.; Methodology, F.Z.G. and A.A.; Software, F.Z.G. and A.A.; Supervision, T.K., A.J. and A.B.; Visualization, H.E.-H.; Writing—original draft, F.Z.G. and A.A. All authors have read and agreed to the published version of the manuscript.

Funding: This research received funding from IRESEN (Institut de Recherche en Energie Solaire et energies Nouvelles).

Institutional Review Board Statement: Not applicable.

Informed Consent Statement: Not applicable.

Data Availability Statement: Data available in a publicly accessible repository.

Acknowledgments: The authors are grateful for the funding support provided by the Institut de Recherche en Energie Solaire et Energies Nouvelles (IRESEN) under the project SOL'R SHEMSY.

Conflicts of Interest: The authors declare no conflict of interest.

References

1. International Energy Agency (IEA). *Energy Policies Beyond IEA Countries: Morocco 2019 Review*; IEA: Paris, France, 2019.
2. Dadush, R.B.U. Has Morocco Benefited from the Free Trade Agreement with the European Union? Policy Center For The New South: Rabat, Morocco, 2020.
3. Bennouna, A.; El Hebil, C. Energy needs for Morocco 2030, as obtained from GDP-energy and GDP-energy intensity correlations. *Energy Policy* **2016**, *88*, 45–55. [CrossRef]
4. Kingdom of Morocco High Comission for Planning. *Growth and Human Development in Morocco Statistical Benchmarks 1998–2008*; Kingdom of Morocco High Comission for Planning: Casablanca, Moroco, 2006.
5. L'environnement, M. *Secteur De L'énergie Chiffres Clés*; 2017.
6. Ghezloun, A.; Saidane, A.; Oucher, N. Energy policy in the context of sustainable development: Case of Morocco and Algeria. *Energy Procedia* **2014**, *50*, 536–543. [CrossRef]
7. Hamdaoui, S.; Mahdaoui, M.; Allouhi, A.; El Alaiji, R.; Kousksou, T.; El Bouardi, A. Energy demand and environmental impact of various construction scenarios of an office building in Morocco. *J. Clean. Prod.* **2018**, *188*, 113–124. [CrossRef]
8. International Energy Agency (IEA). *Energy Prices 2020; High-Quality Data on End-Use Energy Prices*; IEA: Paris, France, 2020.
9. Reboredo, J.C.; Ugolini, A. The impact of energy prices on clean energy stock prices. A multivariate quantile dependence approach. *Energy Econ.* **2018**, *76*, 136–152. [CrossRef]
10. De Gracia, A.; Navarro, L.; Coma, J.; Serrano, S.; Romaní, J.; Pérez, G.; Cabeza, L.F. Experimental set-up for testing active and passive systems for energy savings in buildings–lessons learnt. *Renew. Sustain. Energy Rev.* **2018**, *82*, 1014–1026. [CrossRef]

11. Sun, X.; Gou, Z.; Lau, S.S.-Y. Cost-effectiveness of active and passive design strategies for existing building retrofits in tropical climate: Case study of a zero energy building. *J. Clean. Prod.* **2018**, *183*, 35–45. [CrossRef]

12. Athienitis, A.K.; Barone, G.; Buonomano, A.; Palombo, A. Assessing active and passive effects of façade building integrated photovoltaics/thermal systems: Dynamic modelling and simulation. *Appl. Energy* **2018**, *209*, 355–382. [CrossRef]

13. Addy, M.N.; Adinyira, E.; Ayarkwa, J. Identifying and weighting indicators of building energy efficiency assessment in Ghana. *Energy Procedia* **2017**, *134*, 161–170. [CrossRef]

14. Mahmoud, S.; Zayed, T.; Fahmy, M. Development of sustainability assessment tool for existing buildings. *Sustain. Cities Soc.* **2019**, *44*, 99–119. [CrossRef]

15. Lidelöw, S.; Örn, T.; Luciani, A.; Rizzo, A. Energy-efficiency measures for heritage buildings: A literature review. *Sustain. Cities Soc.* **2019**, *45*, 231–242. [CrossRef]

16. Wang, Y.; Kuckelkorn, J.; Li, D.; Du, J. Evaluation on distributed renewable energy system integrated with a Passive House building using a new energy performance index. *Energy* **2018**, *161*, 81–89. [CrossRef]

17. Bougiatioti, F.; Michael, A. The architectural integration of active solar systems. Building applications in the Eastern Mediterranean region. *Renew. Sustain. Energy Rev.* **2015**, *47*, 966–982. [CrossRef]

18. Vassiliades, C.; Michael, A.; Savvides, A.; Kalogirou, S. Improvement of passive behaviour of existing buildings through the integration of active solar energy systems. *Energy* **2018**, *163*, 1178–1192. [CrossRef]

19. Vieira, A.S.; Stewart, R.A.; Lamberts, R.; Beal, C.D. Residential solar water heaters in Brisbane, Australia: Key performance parameters and indicators. *Renew. Energy* **2018**, *116*, 120–132. [CrossRef]

20. Garnier, C.; Muneer, T.; Currie, J. Numerical and empirical evaluation of a novel building integrated collector storage solar water heater. *Renew. Energy* **2018**, *126*, 281–295. [CrossRef]

21. Chen, X.; Yang, H. Integrated energy performance optimization of a passively designed high-rise residential building in different climatic zones of China. *Appl. Energy* **2018**, *215*, 145–158. [CrossRef]

22. Chen, X.; Yang, H.; Zhang, W. Simulation-based approach to optimize passively designed buildings: A case study on a typical architectural form in hot and humid climates. *Renew. Sustain. Energy Rev.* **2018**, *82*, 1712–1725. [CrossRef]

23. Simona, P.L.; Spiru, P.; Ion, I.V. Increasing the energy efficiency of buildings by thermal insulation. *Energy Procedia* **2017**, *128*, 393–399. [CrossRef]

24. Biswas, K.; Desjarlais, A.; Smith, D.; Letts, J.; Yao, J.; Jiang, T. Development and thermal performance verification of composite insulation boards containing foam-encapsulated vacuum insulation panels. *Appl. Energy* **2018**, *228*, 1159–1172. [CrossRef]

25. Ji, R.; Zhang, Z.; He, Y.; Liu, J.; Qu, S. Simulating the effects of anchors on the thermal performance of building insulation systems. *Energy Build.* **2017**, *140*, 501–507. [CrossRef]

26. Shekar, V.; Krarti, M. Control strategies for dynamic insulation materials applied to commercial buildings. *Energy Build.* **2017**, *154*, 305–320. [CrossRef]

27. Nguyen, D.M.; Grillet, A.-C.; Bui, Q.-B.; Diep, T.M.H.; Woloszyn, M. Building bio-insulation materials based on bamboo powder and bio-binders. *Constr. Build. Mater.* **2018**, *186*, 686–698. [CrossRef]

28. Cetiner, I.; Shea, A.D. Wood waste as an alternative thermal insulation for buildings. *Energy Build.* **2018**, *168*, 374–384. [CrossRef]

29. Aditya, L.; Mahlia, T.; Rismanchi, B.; Ng, H.; Hasan, M.; Metselaar, H.; Muraza, O.; Aditiya, H. A review on insulation materials for energy conservation in buildings. *Renew. Sustain. Energy Rev.* **2017**, *73*, 1352–1365. [CrossRef]

30. Wang, Y.; Jobli, M.I.; Zheng, C.; Yang, X.; Li, K. Thickness of building external insulation in Chongqing based on intermittent heating supply. *Procedia Eng.* **2017**, *205*, 2755–2761. [CrossRef]

31. Küçüktopcu, E.; Cemek, B. A study on environmental impact of insulation thickness of poultry building walls. *Energy* **2018**, *150*, 583–590. [CrossRef]

32. Gao, T.; Ihara, T.; Grynning, S.; Jelle, B.P.; Lien, A.G. Perspective of aerogel glazings in energy efficient buildings. *Build. Environ.* **2016**, *95*, 405–413. [CrossRef]

33. Ghosh, A.; Mallick, T. Evaluation of optical properties and protection factors of a PDLC switchable glazing for low energy building integration. *Sol. Energy Mater. Sol. Cells* **2018**, *176*, 391–396. [CrossRef]

34. Ghosh, A.; Mallick, T. Evaluation of colour properties due to switching behaviour of a PDLC glazing for adaptive building integration. *Renew. Energy* **2018**, *120*, 126–133. [CrossRef]

35. Lu, S.; Li, Z.; Zhao, Q.; Jiang, F. Modified calculation of solar heat gain coefficient in glazing façade buildings. *Energy Procedia* **2017**, *122*, 151–156. [CrossRef]

36. Gao, T.; Jelle, B.P.; Gustavsen, A. *Building Integration of Aerogel Glazings*; Elsevier: Amsterdam, The Netherlands, 2016.

Review

Food, Energy and Water Nexus: A Brief Review of Definitions, Research, and Challenges

Hamdi Abdi [1,*], Maryam Shahbazitabar [1] and Behnam Mohammadi-Ivatloo [2,3]

1   Electrical Engineering Department, Engineering Faculty, Razi University, Kermanshah 67149-67346, Iran; mshahbazitabar@yahoo.com
2   Faculty of Electrical and Computer Engineering, University of Tabriz, Tabriz 51666-16471, Iran; mohammadi@ieee.org
3   Department of Energy Technology, Aalborg University, 9200 Aalborg, Denmark
*   Correspondence: hamdiabdi@gmail.com

Received: 25 September 2020; Accepted: 17 November 2020; Published: 23 November 2020

Abstract: Vast expansion in consumption is leading to natural resource scarcity and global warming. The integrated management of natural resources, such as food, energy, food (FEW) as one of the most important aspects has been proposed as a solution to meet these challenges. The FEW nexus is a world-wide solution for simultaneously assessing the development and implementation of various approaches focusing on energy, water and food security, sufficiency. This approach is intended to foster sustainable development and improve the quality of life of communities while preserving the natural, human and social capital, address the long-term sustainability challenges and protecting all-natural resources. This paper tries to review some recent research on this topic. For this purpose, first, we describe some facts about demand growth and exponential consumption in these three areas, with emphasis on presented statistics. Then, the most critical research published in this field is reviewed, considering that it took a decade or so before that the original idea was introduced. The most important policymakers of this emerging concept, including committees and conferences, and finally significant challenges and opportunities to the implementation along with future insights, are addressed.

Keywords: nexus; food; energy; water; greenhouse gas emission

---

1. Introduction

The food–energy–water (FEW) nexus is known as the significant interconnection between three essential resources for human societies. FEW is an extended concept that is introduced to overcome and manage resource scarceness challenges. The nexus target is to establish effective tradeoffs and synergies between energy, water and food, considering cross-sectoral policies, environmental and social impacts [1]. The FEW nexus implies that the availability of the others may limit constraints or changes of one edge, and implemented solutions in one sector can mainly affect other areas [2,3]. For example, food or energy production is limited by the availability of the water supply. Energy-saving can decrease water consumption, whereas water efficiency growth leads to a decrease in energy, which is used for transition and purification.

As an important note, and based on what is addressed in [4], the perspective of the policymaker determines the method to the FEW nexus. From the perspective of water, energy and food systems are mentioned as the users of the resource, and in the perspective of food and energy, energy and water; water and bio-resources are the outputs, respectively. In addition, each of the three resources affects the others, and ignoring the effects in one resource can have significant impacts on other sources. The main important types of FEW nexus are detailed as energy access and deforestation, biofuels (and unconventional gas and oil) production, hydropower, food and irrigation security and desalinization.

Although many definitions were provided on this concept, the reality is that no comprehensive description has yet been adopted by all researchers [5]. In general terms, the definitions provided for this topic can be divided into two parts.

In the first definition, the nexus is defined as the interactions and interconnections among different sectors (or subsystems) considering food, energy and water [6]. In the second definition, which is more general, the nexus is described as an analysis tool or method to quantify the links among the nexus nodes, including food, energy and water [5].

Over the last few decades, cities are getting more populated than about 54% of the global population [7], and more FEW demands occur. Governments are concerned about FEW sources of scarceness that have effects on several aspects of societies such as community, homes, businesses and industries [8]. This demand growth in cities caused more than 60% of energy consumption and 75% of pollution [9]. In addition, it is estimated food, energy and water resources using will be increased by 35%, 50% and 40%, respectively, by 2030 [10]. The exponential growth of the population considering limited supplies is simulated, and horrible resource scarcities are concluded [11]. Consequently, the increasing rate in population caused more food requests, energy consumption, and increasing greenhouse gas (GHG) emissions because of relying on food chains on water and energy. Although the different natural and fresh products such as wheat, rice, vegetable, etc., need different adequate water by where and when they have cultivated, the canned and frozen food has relied on water supply directly and indirectly.

Water sources are the pillar for both modern and traditional societies. Water is used for irrigation, electricity and heat generation, fuels for transportation and packaging. The authors of [12] implied that total energy utilization in the water section has currently surpassed 800 TW/h, and it is predicted to increase up to 80% until 2040. Indoor agriculture (IA) as a system without farmland is a way to significant energy savings and overcome water wasting (e.g., rooftop greenhouses) [13].

Based on the Energy Information Administration (EIA) reports, the global energy demand will increase by about 50% from 2018 to 2050 [14]. In addition, the energy use in the industrial sector, including mining, refining, agriculture, manufacturing and construction, increases more than 30% between 2018 and 2050 as the demand for goods increases. Furthermore, electricity generation will be increased by 79% between 2018 and 2050. This is while energy generation is a key contributor to air pollutant emissions too. GHG emission, raised by the widespread application of fossil fuels in both the transportation and power generation sector, is the main concern due to environmental issues and the global warming effect. In this way, the climate change raised by GHG emissions, and created by the prevalent fossil fuel usage in both transportation and power generation, is widely accepted as a real-world menace that has potentially severe effects on human health. Hence, one of the critical concerns is moving from fossil fuels to other nonconventional renewable sources of energy, such as wind and solar [15]. As shown in Figure 1, a major contributor to GHG emissions is heat and electricity extraction from oil, coal and natural gas [16]. Even though the nexus concept has been introduced in recent decades, it has evoked the eagerness of researchers all over the world [17].

Preserving the natural resources and preventing their destruction in the three fields of food, energy and water requires a comprehensive framework that considers not only the individual security of the three systems but also their interactions and interdependencies [18]. Figure 2 shows a conceptual model in this regard. As an important note, environmental concerns have become an important security issue in international relations and have been addressed by governments and policymakers. Due to the seriousness of environmental issues and related climate change, this issue can be considered as an effective variable in various conflicts. Climate changes have affected not only the natural habitats but also the social structures of communities and increased the likelihood of violent conflicts [19]. In addition, they endanger international peace and security and affect environmental justice.

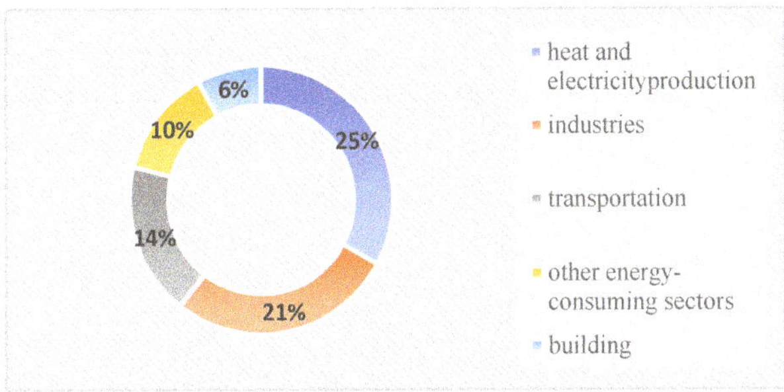

Figure 1. Major contributors to greenhouse gas (GHG) emissions [16].

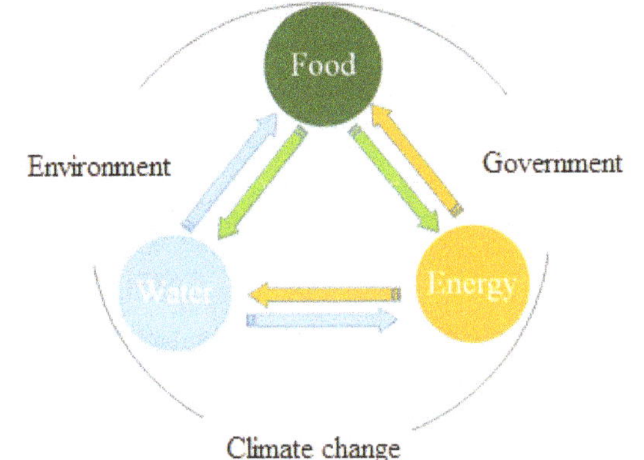

Figure 2. A conceptual model of food–energy–water (FEW) nexus.

Some of the most critical connections between these three systems are as follows:

- Water is needed for energy generation, primarily hydroelectric power plants, biofuels, etc.;
- Water is needed for food production, various nutrients, agricultural irrigation, livestock systems, etc.;
- The energy required for food production, all stages of food preparation including harvesting, transportation, preparation, packaging systems, etc.;
- The energy required in the water sector: water and wastewater purification and desalination, water distribution systems, agricultural irrigation, electricity generation, etc.;
- Nutrition for electricity generation: providing healthy food for personnel and operators in the industrial, economic, etc.

Table 1 provides an example of the energy, food and water security nexus.

Table 1. Energy, food and water nexus [18].

Food Security	Energy Security	Water Security
Food availability	Supply of energy on demand	Water availability
Equal access to food	Physical accessibility of supply	Water health
Optimal water utilization	Supply to satisfy demand at a stable rate	Cost-effectiveness of water

2. Literature Review and Real Case Studies

Since the Bonn conference (2011) with "The Water, Energy and Food Security Nexus—Solutions for the Green Economy" title was held, several numbers of FEW nexus have been published, whereas almost

300 organizations have formed across the world from 2011 to 2015 [17]. The related search keywords were: two-pronged nexus approach "energy–food" (EF), "energy–water" (FW), three-pronged nexus approach "energy–food–water" (EFW) or multipronged such as "energy–food–water–land" (EFWL). In ref. [17], by investigating 37 projects, it was described that 6, 11, 12, and 8 projects (16%, 30%, 32% and 22%) of them had a close linkage with FW, FEW, WE and FEWL nexus, respectively, which is demonstrated in Figure 3.

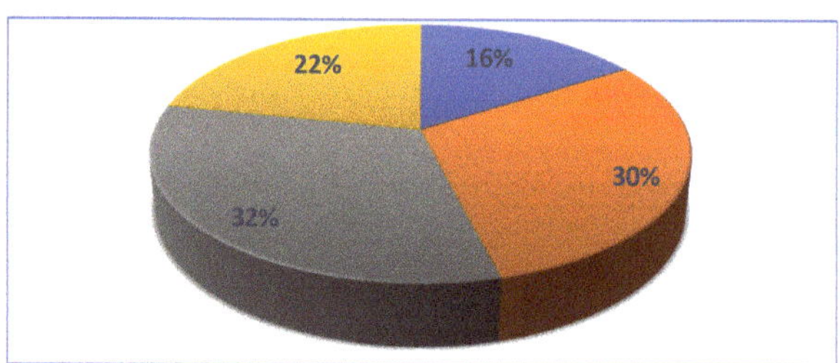

Figure 3. The percentage share of investigated projects in different nexuses in [17].

In another viewpoint, it is claimed every element in this nexus may be a user or source, whereas the taken perspective will affect the policy design. For example, from a water view, energy and food systems are inputs [20], while water and energy are inputs from a food perspective studies [21,22]. Water consumption control considering water supply efficiencies is proposed in food production simultaneously. Projects on energy–water nexus ranged from water for energy to energy for water. Biofuel using water and hydropower generation are examples of water demand for energy production. Energy consumption examples included water pumping in food generation and electricity consumption in water purification.

In [23], a comprehensive survey on modeling and optimization techniques for energy hubs, as one of the basic concepts of future energy systems, is presented. The authors tried to investigate the impacts of distributed energy resources (DERs) in the presence of a smart grid by addressing the economic and environmental considerations. In addition, they mentioned plug-in hybrid electric vehicles (PHEVs), multiple energy flow carriers with storage and the hydrogen economy as a simple EW nexus. Different technologies, including combined heat and power (CHP), solar, wind, hydrogen and different energy vectors, including electrical, natural gas, bio and hydrogen, are addressed in this study.

A sample multi-energy nanogrid (MEN) considering both economic and low-carbon objectives in islanded and grid-connected modes was optimized in [24] by applying linear programming. The case study of a single-family house of 200 m^2 located in Italy was mentioned, and the TRNSYS as the dynamic simulation software is used. The proposed MEN consisted of micro CHP, a boiler and a PV in the generation section, a heat pump and a chiller in the conversion part and some batteries and thermal energy storage (TES) devices for heat and cool in the storage section. This system delivers the electricity, heat and cooling to the end-user.

In [25], a literature review on using nonconventional resources for water and the other alternatives that could be applied to increase food security in water-scarce countries is presented. For this goal, some selected solutions, including the rainfall-runoff water, the desalinated seawater and highly brackish groundwater, marginal-quality water resources and captured by water harvesting, are addressed. In addition, the opportunities contributing to food security in water-scarce countries are detailed among the physical transportation of fresh water and food imports and the "'virtual water" nexus. As a result, this reference claims that there is no choice for water-scarce countries to supply the needed food by applying either the nonconventional or conventional water resources available within their boundaries. In addition, this ref. forecasted that many water-scarce countries should continue to import food. However, they can rely on producing a proportion of their requirements domestically.

Reference [26] classified the main factors to drive food production and agricultural growth in India as access to utilizable water resources and arable land. This reference investigated the water management and food security challenges that lie in the mismatch between agricultural water demand and water availability.

In [27], the training needs and limitations of the oil palm fruit processor in Nigeria by applying a two-stage sampling method to select 160 households included in palm oil extraction activities among the study case are investigated. The data were collected by using an interview and were investigated using both training needs and descriptive analyses.

Both the water needed to run the energy sector and energy applied to the water sector in Spain are investigated in the WE nexus [28]. It considers different policy objectives for expected renewable energy and biofuels resources.

In [29], the role of nonconventional waters (e.g., desalinated and recycled waters), in arid areas, like the city of Kashan, Iran, was focused as the alternative resource of water. This research investigated the costs and environmental impacts of the energy by using these alternative resources. It was claimed that the maximum capacity of the nonconventional water is not necessarily the optimal point.

Different methods of using solar energy for getting usable water from seawater and wastewater were reviewed in [30]. For that goal, the usage of sunlight for water treatment is divided into desalination, or solar distillation, solar detoxification, solar disinfection, and a brief overview of these topics are discussed.

Scott et al. [31] demonstrated the impacts of the coupling of WE at multiple scales. They investigated three WE nexus cases in the US, located in Central, Eastern, and Western regional stakeholder priorities. They stated that localized problems and challenges could be diminished when considered from broader perspectives; meanwhile, regionally significant challenges are not prioritized locally.

In [32], the governing formulation for WE on catchment scales and long-term time are presented in the form of partial differential equations in a three-dimensional space. As the authors claimed, the presented solution is a useful technique to analyze the impacts of land-use and climate changes on the hydrologic cycle.

Hang et al. [33] developed some process systems and engineering tools combined with the nexus concept by using exergy for the design of local generation systems. The suggested framework comprised an optional preliminary design stage combined with a simultaneous design stage based on mathematical optimization. Furthermore, potential interactions between different subsystems were modeled. The suggested method was simulated on a sample WEF at an eco-town in the UK.

The food waste within the FEW nexus is mentioned in [34], and the differential FEW impacts of producing uneaten food and managing food loss and waste were parsed. In addition, different food waste management techniques are classified as landfilling, waste prevention, anaerobic digestion, composting, and incineration. Furthermore, the concept of the "food-waste-systems" technique to optimize resources within the FEW nexus was addressed, and relevant definitions and opportunities were detailed.

The current research needs for understanding the FEW nexus and the relevant implementing solutions in terms of technologies, infrastructures, and policies were discussed in [35]. The main goal of this work is to achieve sustainable development by presenting some guidelines for hydrologists, water resources engineers, economists and policy analysts.

In [36], by focusing on the global value chains (GVCs) concept, the transnational inter-regional input–output approach is used in a tele-connected WEF nexus of East Asia was assessed. For this purpose, China, South Korea and Japan countries were selected as case studies, and different strategies were investigated.

Authors of [37] presented a new framework for optimal short-term scheduling of WE nexus to minimize total electricity cost and seawater desalination considering different constraints. For this purpose, the proposed model was solved by the mixed-integer nonlinear programming (MINLP) using a general algebraic mathematical system (GAMS) software to minimize differed objective functions.

In addition, an interconnected WE nexus, composed of thermal generation units, combined potable water and power (CWP) units, and desalination only processes, is investigated in [38] by using the GAMS software to schedule the water–power hub networks in the presence of the hydro units.

A new technique for the day-ahead optimization of integrated heat–water–electricity systems to optimize different objective functions is proposed in [39], considering a real-time demand-side management strategy. In addition, the impacts of electric vehicle participation are mentioned. The suggested mathematical formulation was solved in the GAMS optimizer using the branch-and-reduce optimization navigator (BARON) tool.

Authors of [40] presented a useful review of demand response strategies applied to the industrial section as well as in the wastewater treatment plant operation.

Peri et al. [41] have investigated the financial impacts of WEF nexus among the volatility spillovers by applying a multivariate generalized autoregressive conditional heteroskedasticity (GARCH) technique. The authors have used the daily data and have applied two indexes of the S&P Goldman Sachs (GS) commodity index for modeling the food and energy variable. In addition, the equity index is used to model the water variable.

In [42], considering an EW nexus, the impact of agriculture and energy prices on the stock performance of the water industry, by applying a multifactor market model, is investigated in a case study that considers the economic and financial crisis during 2008. The results were confirmed in three steps, including a GARCH approach, a rolling OLS and the state space representation estimating Kalman filter.

In addition, modeling the possible interactions between the future energy prices in European Union Allowances (EUA) is addressed in [43] by applying a nonlinear co-integration dynamic method through CO_2 futures and Brent.

Table 2 reviews some research information on this context.

Table 2. Brief information about studies in different nexus.

Ref No.	Nexus	Aspects of Goal	Description	Year
[23]	WE	A review on economic, environmental and emission issues of EH	Different technologies and energy carriers are investigated	2019
[24]	WE	Economic and Environmental	Micro CHP, boiler, PV, heat pump, chiller batteries and TES devices	2020
[25]	WF	Environmental	Nonconventional water sources for achieving food security in arid countries are used	2007
[26]	WF	Environmental, social, economical	Pro-rata pricing in farm and public irrigation systems improves the energy efficiency in green water; the residual soil moisture depletion preventing; low water consuming crops cultivation	2012
[27]	WF	Social, economical	Expansion and training studies by stakeholders for palm oil extraction in Nigeria	2011
[28]	WE	Social, economical	More efficient ways for irrigation, urban wastewater menace and the use of desalinated water	2012
[29]	WE	Economical	Total water shortage can be compensated by increasing the production of nonconventional water	2019
[30]	WE	Economical	Using sunlight for purification of water	2019

Table 2. *Cont.*

Ref No.	Nexus	Aspects of Goal	Description	Year
[31]	WE	Environmental, political	Addressing the methods for coupling different resources at multiple scales, find out the obstacle of institutional opportunities for decision-making	2011
[32]	WE	Economical	A unique general solution for the mean annual energy–water balance equation has been proposed by using mathematical formulation and dimensional analysis	2008
[33]	FEW	Economical	Local production system method and extract optimization model for the energy–food–water nexus designing in a local UK eco-town.	2016
[34]	FEW	Environmental, social, economical	Addressing the different subsystems for preventable and unpreventable food waste	2018
[35]	FEW	Economic, political	Historical efforts in integrated water resources management (IWRM) have been applied to present alternatives for interdisciplinary studies among several groups with collaboration between food and energy communities.	2018
[36]	FEW	Environmental, social, political, economical	Designing the hybrid FEW connections in Japan and China to obtain the interdependencies of hybrid water, hybrid energy and food extractions with other sectors in two countries	2019
[37]	WE	Economic dispatch (ED)	Day-ahead ED, including coupled desalinated water, power networks in the presence of compressed air energy storages	2019
[38]	WE	Short-term scheduling	Short-term planning of desalination water and thermal units	2019
[39]	WE	Short-term scheduling	Investigating the impacts of demand response programs and plug-in electric vehicles in short term scheduling of a heat–energy–power system	2019
[40]	WE	Demand response	A review on demand response in energy–water nexus	2019
[41]	WEF	Financial impact of nexus	An investigation of volatility spillover in Europe, Asia, North America, Latin America and the world is addressed	2017
[42]	WE	Financial impacts of nexus	Analyzing the impact of agriculture and energy prices on the water industry	2018
[43]	E	Financial	Modeling future energy prices in EUA	2011
[44]	WF	Environmental, social, economical	The microfinance funding model, public–private cooperation, using data-intensive methods such as climate forecasting models for agriculture	2015

In addition, due to the great importance of this novel subject, different techniques have been adapted around the world on real case studies or projects to prove the effectiveness of this subject. Some of the significant studies are highlighted in Table 3.

Table 3. Brief data on some highlighted real case studies on different nexus.

Ref No.	Nexus	Aspects of Goal	Location	Year
[45]	WE	Recovery of energy from wastewater treatment plants	United States	2010
[46]	WE	Energy consumption	Bangladesh	2012
[47]	FEW	The impacts of nexus on tourism	The Mediterranean Region	2014
[48]	FEW	The impacts on nexus on transboundary context	The Euphrates–Tigris river basin	2015
[49]	FEW	Transboundary river	Tonle Sap Lake, Mekong River Basin	2015
[50]	WE	Urban agglomeration based on multiregional data	Beijing–Tianjin–Hebei region, China	2016
[51]	FEW	The urban systems fundamental to investigate the transboundary FEW	Delhi, India	2017
[52]	WE	Proposing the reference resource-to-service system framework	New York City	2017
[53]	FEW	System analysis and interactive visualization	The Great Ruaha River of Tanzania	2018
[54]	WE	A review of tools and methods or assessment of macro WE nexus is presented	70 case studies over the world are surveyed, and 35 comprehensive macro-level case studies are detailed in levels of the city, regional, national, transboundary	2018
[55]	WE	Investigating the construction industry	China's at the provincial level	2019
[56]	Climate, land, energy and water (CLEW) nexus	Analyzing the energy sustainability challenges	Lebanon	2019
[57]	FEW	Assessment of nexus by applying a decision support technique	Saudi Arabia	2019
[58]	FEW	Investigating some direct and indirect nexus at metropolitan statistical areas	United States	2019
[59]	FEW	Applying the stochastic multicriteria decision-making (MCDM) technique for investigating the desirability of different energy generation methods	Indonesia	2019
[60]	FEW	Presenting a toolbox for interactive analysis	At the country-level, for specified categories	2020

3. Nexus Committee, Conferences and Real Case Studies

Various conferences, committees, projects and research highlights were organized around the world, relying on nexus. Table 4 shows brief information about the committee and gathering.

Table 4. Committee and conference information about nexus.

Name	Year	Title	Subject	Location
United Nations University (UNU)	1983	Food–Energy Nexus Program	Food–energy interconnections	Brazil
United Nations University (UNU)	1984	Food, Energy and Ecosystems	Food–energy interconnections	Brazil
United Nations University (UNU)	1986	Food–Energy and Ecosystems	Energy consumption patterns and their effects on ecosystem and agriculture	India
World Bank	The 1990s	Water, food and trade		
Columbia Water Center of the Earth Institute at Columbia University	2000	Water–energy–agriculture	Water and climate interact with food, energy, ecosystems and urbanization	India
Kyoto World Water Forum	2003	Virtual water	Water as a pillar in the nexus	japan
Bonn Nexus Conference	2011	Water, energy and food security nexus, Solutions for the green economy	Nexus challenge, Increase policy coherence, end waste and minimize losses	Germany
Rio+20	2012	Green economy	Political outcome, Sustainable Development	Brazil
United Nations University Institute for Integrated Management of Material Fluxes and Resources (UNU-FLORES)	2012	Water, waste and soil	Interdependencies of environmental resources and the interconnection between compartments.	
United Nations Economic and Social Commission for Asia and the Pacific (UN-ESCAP)	2013	Food–Energy–Water	Water–energy–food nexus, synergies and tradeoffs	Asia and the Pacific
Food and Agriculture Organization (FAO)	2013	Food–Energy–Water	International efforts to defeat hunger and improve local economies	
Bonn Nexus Conference	2014	Sustainability in the water–energy–food nexus	Financial, institutional, technical and intellectual resource development for nexus research and applications	Germany

4. Nexus Challenges

Given the broad scope of the FEW nexus, including its three large-scale systems, as well as its newness and capitalization of its wide-ranging discussions, there are significant challenges to its implementation. Different researchers have focused on some of these challenges, as well as some relevant solutions. Some of these items are addressed as follows:

Authors of [61] reflected recent research in stakeholder engagement regarding the nexus field. The paper outlined four main emerging concepts in this regard, including using the trans-disciplinary approaches for assessing as well as visualizing nexus, the understanding capacity of building and governance, accounting for inter-scalar and multi-relationships and investigating the implications of future socioeconomic, climatic and technological changes. The authors claimed that "it seems likely that without new trans-disciplinary approaches, there will continue to be poor coordination in addressing challenges across the water food and energy domains".

The existence of some serious challenges in the issues of planning, operation and privatization in the current three separate subsystems of food, water, and energy and necessitate to present the practical, useful and economical solutions to these issues in the FEW nexus; needing some new laws and governance structures and mechanisms for implementing the nexus concept, needing to find some solutions to the consequences of implementing this integrated system, are the most challenging issues from the perspective of reference [62].

Zaidi et al. [63] addressed the water–energy nexus problems and alternatives focusing on machine learning contexts. They categorized various challenges as data challenges: missing data, spatiotemporal data, heterogeneity in data, data collection standards and data availability; and machine learning challenges: modeling spatiotemporal data, modeling in the presence of missing data, identifying outliers (including imperfect collection methods/sensors and extreme events). In addition, they categorized the machine learning techniques used in the energy–water nexus based on applying artificial neural networks (ANN), support vector machines (SVM), time–series analysis, regression, unsupervised learning (including Bayesian model averaging, random forests and hybrid models) and reinforcement learning in energy generation, energy use, water use, energy for water and water for energy. Furthermore, the authors detailed the machine learning alternatives for the energy–water nexus as mining patterns and relationships in data, addressing heterogeneity in data, predicting energy–water nexus variables, modeling unobserved variables, integration of models and deep learning.

Reference [64] outlined the directions for researchers, decision-makers or practitioners, as Identifying nexus methods as well as tools which are suitable for implementation in nexus approaches, creating a practical roadmap, distinguishing the targeted resources, structures and subsystems, deciding system constraints and boundaries, categorizing method types based on research objectives, determining the capabilities and drawbacks and limitations of existing techniques and tools, classifying bottom–up and top–down techniques, describing the uncertain data and uncertain modeling. Authors of [5] categorized the future studies in the nexus field as system boundary, data and modeling uncertainty, the essential mechanism of the nexus and the evaluation of coupled nexus systems even though internal and external impact analysis is of great importance in this regard.

In addition, introducing new concepts such as a smart city (SC), mainly in power systems, has forced the operators and planners to search for new secure and reliable solutions for implementing the Nexus missions. In this regard, the reliability and security of design, planning and operation concepts of different nexuses in SCs should be analyzed and investigated carefully [65].

Analyzing various references and studies related to the challenges of nexus systems, which some of them are mentioned above, shows that there are some significant challenges, as well as solutions in implementing the nexus systems. Table 5 summarizes some of these concepts.

Table 5. Some of the challenges and solutions for implementing nexus systems.

Challenges	Solutions
Lack of integrated policy and legislation for the system	Integrated policy-making such as integrated pricing in water and energy fields, developing a model of agricultural complex and industry proper allocation
Data uncertainty	Implementing the appropriate uncertainty modeling such as stochastic programming, scenario generation, and so on.
Large numbers of data for subsystems	Applying data-mining techniques
System boundary	Accurate detection of cases using precise and rapid identification of subsystems
Lack of sufficient standards and laws	Forming committees comprising subdiscipline specialists to address this gap
Lack of efficient software platforms	Presenting multi-domain software

5. Conclusions

In this paper, an overview of the processes, methods, policies and interconnections of several resources was reviewed. Recent researches show the trend of societies to establish a comprehensive plan for handling energy, food and water consumption effectively. This paper implies a different nexus method relationship with inherent water, energy and food resource interactions approach considering political and environmental aspects. We reviewed the current state of research on nexus approaches. Food, energy and water systems are interconnected in such a way that one action in a system often affects the others. Therefore, centralized techniques to investigate, planning and operation should be integrated to reduce the side-effects and increase collaboration and synergism. A significant interconnection between these three subsystems and their direct impacts on environmental concepts, climate changes, socioeconomic, policy-making, etc., needs stakeholder engagement so that integrated management overall subsystems is concerned. This is essential for achieving the nexus objectives and gaining sustainable development. Planning and policy-making between the departments and organizations implemented to get a common point to require the discourse between stakeholders and the structures of conflicting objectives to fully cooperate and reduce the interference. As it was concluded, the main challenges for implementing this concept are categorized as lack of integrated policy and legislation for the system, data uncertainty and large numbers data of subsystems, system boundary, lack of sufficient standards and laws and lack of efficient software platforms.

Author Contributions: H.A.: Main idea, data collecting, classifications, conceptualization, methodology, wtitting the original and revised versios; M.S.: Conceptualization, methodology, wtitting the original and revised versios; B.M.-I.: Conceptualization, methodology, wtitting the original and revised versios. All authors have read and agreed to the published version of the manuscript.

Funding: This research received no external funding.

Conflicts of Interest: The authors declare no conflict of interest.

References

1. Albrecht, T.R.; Crootof, A.; Scott, C.A. The Water-Energy-Food Nexus: A systematic review of methods for nexus assessment. *Environ. Res. Lett.* **2018**, *13*, 043002. [CrossRef]
2. Griffith, D.; Johnson, D.; Hunt, A. The geographic distribution of metals in urban soils: The case of Syracuse, NY. *GeoJournal* **2009**, *74*, 275–291. [CrossRef]
3. Eftelioglu, E.; Jiang, Z.; Tang, X.; Shekhar, S. The nexus of food, energy, and water resources: Visions and challenges in spatial computing. In *Advances in Geocomputation*; Springer: Cham, Switzerland, 2017; pp. 5–20.
4. Bazilian, M.; Rogner, H.; Howells, M.; Hermann, S.; Arent, D.; Gielen, D.; Steduto, P.; Mueller, A.; Komor, P.; Tol, R.S. Considering the energy, water and food nexus: Towards an integrated modelling approach. *Energy Policy* **2011**, *39*, 7896–7906. [CrossRef]
5. Zhang, C.; Chen, X.; Li, Y.; Ding, W.; Fu, G. Water-energy-food nexus: Concepts, questions and methodologies. *J. Clean. Prod.* **2018**, *195*, 625–639. [CrossRef]
6. Sanders, K.T.; Webber, M.E. Evaluating the energy consumed for water use in the United States. *Environ. Res. Lett.* **2012**, *7*, 034034. [CrossRef]
7. Gutjahr, W.J. Convergence Analysis of Metaheuristics. In *Matheuristics*; Springer: Boston, MA, USA, 2009; pp. 159–187.
8. Bavafa, M.; Navidi, N.; Monsef, H. A new approach for Profit-Based Unit Commitment using Lagrangian relaxation combined with ant colony search algorithm. In Proceedings of the 2008 43rd International Universities Power Engineering Conference, Padova, Italy, 1–4 September 2008.
9. Rampriya, B.; Mahadevan, K. Scheduling the units and maximizing the profit of GENCOS using LR-PSO technique. *Int. J. Electr. Eng. Inform.* **2010**, *2*, 150–158. [CrossRef]
10. National Intelligence Council. *Global Trends 2030: Alternative Worlds: A Publication of the National Intelligence Council*; U.S. Government Printing Office: Washington, DC, USA, 2012.
11. Meadows, D.H.; Meadows, D.L.; Randers, J.; Behrens, W.W., III. *The Limits to Growth: A Report for the Club of Rome's Project on the Predicament of Mankind*; Universe Books: New York, NY, USA, 1972.

12. Brouwer, F.; Avgerinopoulos, G.; Fazekas, D.; Laspidou, C.; Mercure, J.-F.; Pollitt, H.; Ramos, E.P.; Howells, M. Energy modelling and the Nexus concept. *Energy Strategy Rev.* **2018**, *19*, 1–6. [CrossRef]

13. Specht, K.; Siebert, R.; Hartmann, I.; Freisinger, U.B.; Sawicka, M.; Werner, A.; Thomaier, S.; Henckel, D.; Walk, H.; Dierich, A. Urban agriculture of the future: An overview of sustainability aspects of food production in and on buildings. *Agric. Hum. Values* **2014**, *31*, 33–51. [CrossRef]

14. Global Enerzgy Consumption to Increase by 50 by 2050. Available online: https://safety4sea.com/global-energy-consumption-to-increase-by-50-by-2050/ (accessed on 18 November 2020).

15. Shahbazitabar, M.; Abdi, H. A novel priority-based stochastic unit commitment considering renewable energy sources and parking lot cooperation. *Energy* **2018**, *161*, 308–324. [CrossRef]

16. Wang, X.-C.; Klemeš, J.J.; Dong, X.; Fan, W.; Xu, Z.; Wang, Y.; Varbanov, P.S. Air pollution terrain nexus: A review considering energy generation and consumption. *Renew. Sustain. Energy Rev.* **2019**, *105*, 71–85. [CrossRef]

17. Endo, A.; Tsurita, I.; Burnett, K.; Orencio, P.M. A review of the current state of research on the water, energy, and food nexus. *J. Hydrol. Reg. Stud.* **2017**, *11*, 20–30. [CrossRef]

18. Bizikova, L.; Roy, D.; Venema, H.D.; McCandless, M.; Swanson, D.; Khachtryan, A.; Borden, C.; Zubrycki, K. *Water-Energy-Food Nexus and Agricultural Investment: A Sustainable Development Guidebook*; International Institute for Sustainable Development (IISD): Winnipeg, MB, Canada, 2014.

19. AKHAVAN KAZEMI, M.; Sadat Hoseini, T.; Bahramipoor, F. Analysis of the Impact of Climate Change on International Security. *Res. Lett. Int. Relat.* **2019**, *12*, 9–39.

20. Hellegers, P.; Zilberman, D.; Steduto, P.; McCornick, P. Interactions between water, energy, food and environment: Evolving perspectives and policy issues. *Water Policy* **2008**, *10*, 1–10. [CrossRef]

21. Mushtaq, S.; Maraseni, T.N.; Maroulis, J.; Hafeez, M. Energy and water tradeoffs in enhancing food security: A selective international assessment. *Energy Policy* **2009**, *37*, 3635–3644. [CrossRef]

22. Khan, S.; Hanjra, M.A. Footprints of water and energy inputs in food production–Global perspectives. *Food Policy* **2009**, *34*, 130–140. [CrossRef]

23. Maroufmashat, A.; Taqvi, S.T.; Miragha, A.; Fowler, M.; Elkamel, A. Modeling and Optimization of Energy Hubs: A Comprehensive Review. *Inventions* **2019**, *4*, 50. [CrossRef]

24. Di Somma, M.; Caliano, M.; Graditi, G.; Pinnarelli, A.; Menniti, D.; Sorrentino, N.; Barone, G. Designing of cost-effective and low-carbon multi-energy nanogrids for residential applications. *Inventions* **2020**, *5*, 7. [CrossRef]

25. Qadir, M.; Sharma, B.R.; Bruggeman, A.; Choukr-Allah, R.; Karajeh, F. Non-conventional water resources and opportunities for water augmentation to achieve food security in water scarce countries. *Agric. Water Manag.* **2007**, *87*, 2–22. [CrossRef]

26. Kumar, M.D.; Sivamohan, M.; Narayanamoorthy, A. The food security challenge of the food-land-water nexus in India. *Food Secur.* **2012**, *4*, 539–556. [CrossRef]

27. Akangbe, J.; Adesiji, G.; Fakayode, S.; Aderibigbe, Y. Towards palm oil self-sufficiency in Nigeria: Constraints and training needs nexus of palm oil extractors. *J. Hum. Ecol.* **2011**, *33*, 139–145. [CrossRef]

28. Hardy, L.; Garrido, A.; Juana, L. Evaluation of Spain's water-energy nexus. *Int. J. Water Resour. Dev.* **2012**, *28*, 151–170. [CrossRef]

29. Noruzi, M.; Yazdandoost, F. Determining the Optimal Point In Arid Basins Using Water-Energy Nexus Approach. *Int. J. Optim. Civ. Eng.* **2019**, *9*, 423–435.

30. Chakraborty, S. WATER–ENERGY NEXUS: ROLE OF SOLAR ENERGY. *PREPARE@ U-Preprint Archive*, 2019. [CrossRef]

31. Scott, C.A.; Pierce, S.A.; Pasqualetti, M.J.; Jones, A.L.; Montz, B.E.; Hoover, J.H. Policy and institutional dimensions of the water–energy nexus. *Energy Policy* **2011**, *39*, 6622–6630. [CrossRef]

32. Yang, H.; Yang, D.; Lei, Z.; Sun, F. New analytical derivation of the mean annual water-energy balance equation. *Water Resour. Res.* **2008**, *44*. [CrossRef]

33. Hang, M.Y.L.P.; Martinez-Hernandez, E.; Leach, M.; Yang, A. Designing integrated local production systems: A study on the food-energy-water nexus. *J. Clean. Prod.* **2016**, *135*, 1065–1084. [CrossRef]

34. Kibler, K.M.; Reinhart, D.; Hawkins, C.; Motlagh, A.M.; Wright, J. Food waste and the food-energy-water nexus: A review of food waste management alternatives. *Waste Manag.* **2018**, *74*, 52–62. [CrossRef]

35. Cai, X.; Wallington, K.; Shafiee-Jood, M.; Marston, L. Understanding and managing the food-energy-water nexus–opportunities for water resources research. *Adv. Water Resour.* **2018**, *111*, 259–273. [CrossRef]

36. White, D.J.; Hubacek, K.; Feng, K.; Sun, L.; Meng, B. The Water-Energy-Food Nexus in East Asia: A teleconnected value chain analysis using inter-regional input-output analysis. *Appl. Energy* **2018**, *210*, 550–567. [CrossRef]

37. Jabari, F.; Mohammadi Ivatloo, B.; Sharifian, B.; Ghaebi, H. Day-ahead economic dispatch of coupled desalinated water and power grids with participation of compressed air energy storages. *J. Oper. Autom. Power Eng.* **2019**, *7*, 40–48.

38. Jabari, F.; Mohammadi Ivatloo, B.; Sharifian, B.; Ghaebi, H. Optimal Short-Term Coordination of Desalination, Hydro and Thermal Units. *J. Oper. Autom. Power Eng.* **2019**, *7*, 141–147.

39. Jabari, F.; Jabari, H.; Mohammadi-ivatloo, B.; Ghafouri, J. Optimal short-term coordination of water-heat-power nexus incorporating plug-in electric vehicles and real-time demand response programs. *Energy* **2019**, *174*, 708–723. [CrossRef]

40. Kirchem, D.; Lynch, M.Á.; Bertsch, V.; Casey, E. Modelling demand response with process models and energy systems models: Potential applications for wastewater treatment within the energy-water nexus. *Appl. Energy* **2020**, *260*, 114321. [CrossRef]

41. Peri, M.; Vandone, D.; Baldi, L. Volatility spillover between water, energy and food. *Sustainability* **2017**, *9*, 1071. [CrossRef]

42. Vandone, D.; Peri, M.; Baldi, L.; Tanda, A. The impact of energy and agriculture prices on the stock performance of the water industry. *Water Resour. Econ.* **2018**, *23*, 14–27. [CrossRef]

43. Peri, M.; Baldi, L. Nonlinear price dynamics between CO2 futures and Brent. *Appl. Econ. Lett.* **2011**, *18*, 1207–1211. [CrossRef]

44. Govardhan, M.; Mishra, M.; Sundeep, S.; Roy, R. Solution of price based unit commitment using GABC and TLBO optimization algorithms. In Proceedings of the 2014 International Conference on Control, Instrumentation, Energy and Communication (CIEC), Calcutta, India, 31 January–2 February 2014.

45. Stillwell, A.S.; Hoppock, D.C.; Webber, M.E. Energy recovery from wastewater treatment plants in the United States: A case study of the energy-water nexus. *Sustainability* **2010**, *2*, 945–962. [CrossRef]

46. Alam, M.J.; Begum, I.A.; Buysse, J.; Van Huylenbroeck, G. Energy consumption, carbon emissions and economic growth nexus in Bangladesh: Cointegration and dynamic causality analysis. *Energy Policy* **2012**, *45*, 217–225. [CrossRef]

47. Tugcu, C.T. Tourism and economic growth nexus revisited: A panel causality analysis for the case of the Mediterranean Region. *Tour. Manag.* **2014**, *42*, 207–212. [CrossRef]

48. Kibaroglu, A.; Gürsoy, S.I. Water–energy–food nexus in a transboundary context: The Euphrates–Tigris river basin as a case study. *Water Int.* **2015**, *40*, 824–838. [CrossRef]

49. Keskinen, M.; Someth, P.; Salmivaara, A.; Kummu, M. Water-energy-food nexus in a transboundary river basin: The case of Tonle Sap Lake, Mekong River Basin. *Water* **2015**, *7*, 5416–5436. [CrossRef]

50. Wang, S.; Chen, B. Energy–water nexus of urban agglomeration based on multiregional input–output tables and ecological network analysis: A case study of the Beijing–Tianjin–Hebei region. *Appl. Energy* **2016**, *178*, 773–783. [CrossRef]

51. Ramaswami, A.; Boyer, D.; Nagpure, A.S.; Fang, A.; Bogra, S.; Bakshi, B.; Cohen, E.; Rao-Ghorpade, A. An urban systems framework to assess the trans-boundary food-energy-water nexus: Implementation in Delhi, India. *Environ. Res. Lett.* **2017**, *12*, 025008. [CrossRef]

52. Engström, R.E.; Howells, M.; Destouni, G.; Bhatt, V.; Bazilian, M.; Rogner, H.-H. Connecting the resource nexus to basic urban service provision–with a focus on water-energy interactions in New York City. *Sustain. Cities Soc.* **2017**, *31*, 83–94. [CrossRef]

53. Yang, Y.E.; Wi, S. Informing regional water-energy-food nexus with system analysis and interactive visualization—A case study in the Great Ruaha River of Tanzania. *Agric. Water Manag.* **2018**, *196*, 75–86. [CrossRef]

54. Dai, J.; Wu, S.; Han, G.; Weinberg, J.; Xie, X.; Wu, X.; Song, X.; Jia, B.; Xue, W.; Yang, Q. Water-energy nexus: A review of methods and tools for macro-assessment. *Appl. Energy* **2018**, *210*, 393–408. [CrossRef]

55. Hong, J.; Zhong, X.; Guo, S.; Liu, G.; Shen, G.Q.; Yu, T. Water-energy nexus and its efficiency in China's construction industry: Evidence from province-level data. *Sustain. Cities Soc.* **2019**, *48*, 101557. [CrossRef]

56. Abou Farhat, R.; Mahlooji, M.; Gaudard, L.; El-Baba, J.; Harajli, H.; Kabakian, V.; Madani, K. A Multi-attribute Assessment of Electricity Supply Options in Lebanon. In *Food-Energy-Water Nexus Resilience and Sustainable Development*; Springer: Cham, Switzerland, 2020; pp. 1–27.

57. Vittorio, M. A Decision Support Tool for the Assessment of Water–Energy–Food Nexus in Saudi Arabia. In *Food-Energy-Water Nexus Resilience and Sustainable Development*; Springer: Cham, Switzerland, 2020; pp. 57–73.

58. Djehdian, L.A.; Chini, C.M.; Marston, L.; Konar, M.; Stillwell, A.S. Exposure of urban food–energy–water (FEW) systems to water scarcity. *Sustain. Cities Soc.* **2019**, *50*, 101621. [CrossRef]

59. Mahlooji, M.; FGumilar, G.; Madani, K. Dealing with Trade-offs in Sustainable Energy Planning: Insight for Indonesia. In *Food-Energy-Water Nexus Resilience and Sustainable Development*; Springer: Cham, Switzerland, 2020; pp. 243–266.

60. Sadegh, M.; AghaKouchak, A.; Mallakpour, I.; Huning, L.S.; Mazdiyasni, O.; Niknejad, M.; Foufoula-Georgiou, E.; Moore, F.C.; Brouwer, J.; Farid, A. Data and analysis toolbox for modeling the nexus of food, energy, and water. *Sustain. Cities Soc.* **2020**, *61*, 102281. [CrossRef]

61. Hoolohan, C.; Larkin, A.; McLachlan, C.; Falconer, R.; Soutar, I.; Suckling, J.; Varga, L.; Haltas, I.; Druckman, A.; Lumbroso, D. Engaging stakeholders in research to address water–energy–food (WEF) nexus challenges. *Sustain. Sci.* **2018**, *13*, 1415–1426. [CrossRef]

62. Larson, R.B.; Holley, C.; Bowman, D.M. THE ENERGY/WATER/FOOD NEXUS—AN INTRODUCTION. *Jurimetr. J. Lawscience Technol.* **2018**, *59*, 1–14.

63. Zaidi, S.M.A.; Chandola, V.; Allen, M.R.; Sanyal, J.; Stewart, R.N.; Bhaduri, B.L.; McManamay, R.A. Machine learning for energy-water nexus: Challenges and opportunities. *Big Earth Data* **2018**, *2*, 228–267. [CrossRef]

64. Endo, A.; Yamada, M.; Miyashita, Y.; Sugimoto, R.; Ishii, A.; Nishijima, J.; Fujii, M.; Kato, T.; Hamamoto, H.; Kimura, M. Dynamics of water–energy–food nexus methodology, methods, and tools. *Curr. Opin. Environ. Sci. Health* **2020**, *13*, 46–60. [CrossRef]

65. Abdi, H.; Shahbazitabar, M. Smart city: A review on concepts, definitions, standards, experiments, and challenges. *J. Energy Manag. Technol.* **2020**, *4*, 1–6.

Publisher's Note: MDPI stays neutral with regard to jurisdictional claims in published maps and institutional affiliations.

MDPI

St. Alban-Anlage 66

4052 Basel

Switzerland

Tel. +41 61 683 77 34

Fax +41 61 302 89 18

www.mdpi.com

Inventions Editorial Office

E-mail: inventions@mdpi.com

www.mdpi.com/journal/inventions